D0881040

Telecommunications Network Management

IEEE Press
445 Hoes Lane, P.O. Box 1331
Piscataway, NJ 08855-1331

Editorial Board
Roger F. Hoyt, *Editor in Chief*

J. B. Anderson	A. H. Haddad	R. S. Muller
P. M. Anderson	R. Herrick	W. D. Reeve
M. Eden	G. F. Hoffnagle	D. J. Wells
M. E. El-Hawary	S. Kartalopoulos	
S. Furui	P. Laplante	

Kenneth Moore, *Director of IEEE Press*
Karen Hawkins, *Senior Acquisitions Editor*
Linda Matarazzo, *Editorial Assistant*
Surendra Bhimani, *Production Editor*

IEEE Communications Society, *Sponsor*
COMM-S Liaison to IEEE Press, Salah Aidarous

Cover Design: William T. Donnelly, *WT Design*

Technical Reviewers

Mr. Thomas Grim, *Southwestern Bell*
Professor Thomas Robertazzi, *State University of New York at Stony Brook*

Books of Related Interest from IEEE Press . . .

TELECOMMUNICATION NETWORK MANAGEMENT INTO THE 21ST CENTURY
Techniques, Standards, Technologies, and Applications
Salah Aidarous and Thomas Plevyak
1996   Hardcover   448pp   IEEE Order No. PC3624   ISBN 0-7803-1013-6

ENGINEERING NETWORKS FOR SYNCHRONIZATION, CCS 7, AND ISDN
Standards, Protocols, Planning, and Testing
P. K. Bhatnagar
1997   Hardcover   512pp   IEEE Order No. PC5628   ISBN 0-7803-1158-2

SUBSCRIBER LOOP SIGNALING AND TRANSMISSION HANDBOOK: Analog
Whitham D. Reeve
1992   Hardcover   672pp   IEEE Order No. PC3376   ISBN 0-7803-0440-3

SUBSCRIBER LOOP SIGNALING AND TRANSMISSION HANDBOOK: Digital
Whitham D. Reeve
1995   Hardcover   672 pp   IEEE Order No. PC3376   ISBN 0-7803-0440-3

COMMUNICATIONS SYSTEMS AND TECHNIQUES: An IEEE Press Classic Reissue
Mischa Schwartz, William R. Bennett, and Seymour Stein
1966   Hardcover   692pp   IEEE Order No. PC5639   ISBN 0-7803-11-66-3

Telecommunications Network Management

Technologies and Implementations

Edited by

Salah Aidarous
NEC America

Thomas Plevyak
Bell Atlantic

 **IEEE PRESS**

IEEE Series on Network Management
Salah Aidarous and Thomas Plevyak, *Series Editors*

IEEE Communications Society, *Sponsor*

The Institute of Electrical
and Electronics Engineers, Inc., New York

This book and other books may be purchased at a discount
from the publisher when ordered in bulk quantities. Contact:

IEEE Press Marketing
Attn: Special Sales
Piscataway, NJ 08855-1331
Fax: (732) 981-9334

For more information about IEEE PRESS products,
visit the IEEE Home Page: http://www.ieee.org/

©1998 by the Institute of Electrical and Electronics Engineers, Inc.
345 East 47th Street, New York, NY 10017-2394

All rights reserved. No part of this book may be reproduced in any form,
nor may it be stored in a retrieval system or transmitted in any form,
without written permission from the publisher.

Printed in the United States of America

10   9   8   7   6   5   4   3   2   1

ISBN 0-7803-3454-X
IEEE Order Number: PC5711

Library of Congress Cataloging-in-Publication Data
Telecommunications network management : technologies
  and implementations / edited by Salah Aidarous, Thomas Plevyak.
     p. cm. – (IEEE series on network management)
    "IEEE Communications Society, sponsor."
    Includes bibliographical references and index.
    ISBN 0-7803-3454-X
    1. Telecommunications systems–Management. 2. Computer networks–
Management. I. Aidarous, Salah. II. Plevyak, Thomas. III. IEEE
Communications Society.
IV. Series.
TK5102.5.T4234 1997
621.382″1–dc21   97–38066
                                                                      CIP

The Editors and Authors dedicate this book to their colleagues around the world whose contributions are advancing the critically important field of network management.

Contents

Chapter 2   **OSI Systems Management, Internet SNMP, and
ODP/OMG CORBA as Technologies for Telecommunications
Network Management 63**
George Pavlou, University College London, UK

Chapter 3   **Management Platforms 111**
George Pauthner, Alcatel Telecom, Germany
Jerry Power, Alcatel Telecom, USA

Maurizio Decina
Politecnico di Milano/CEFRIEL

Guest Introduction

*TMN Today: Challenges
and Opportunities*

The whole business of telecommunications is undergoing a period of dramatic change. The opening of most advanced markets to competition and the concurrent globalization process have led to a dynamic and unpredictable environment, where incumbent network operators are facing the challenge posed by new local entrants and by international "global" network operators. The skills needed to be successful are: ability to understand and anticipate customer needs, quickness in adapting the service portfolio to market requirements, and capability to offer services at competitive prices.

On the technology side, the pace of change is always increasing, turning the advanced development of last year into the affordable commodity of today. Moreover, a major convergence process between telecommunications, consumer electronics, and computing is now taking place, giving rise to a new market whose superb potential is still hard to predict and estimate.

All these changes can be reduced to a common denominator, which is the change of the very nature of the telecommunications industry from a "technology-driven" to a "market-driven" industry. This evolution is going to modify, forever, the way of doing business in this field. In the past, the main competitive advantage was the technical superiority of network products and the performance guaranteed to basic services. Such characteristics are typical of monopoly markets at an early stage of evolution. The present transition is toward a more mature scenario where competition skills are increasingly based on aspects such as service quality, price negotiation, and customer care support.

Perhaps the area where implications of these changes are most evident is the *Telecommunications Management Network* (TMN). Early network management systems addressed the ability to operate large and complex networks, and focused on the basic *Operations, Administration, and Maintenance* (OAM) aspects: configuring the network, ensuring its availability, and collecting basic performance and accounting data. The majority of these features are tightly coupled to the operators' internal processes and are almost invisible to the external world and to customers in particular.

All these aspects, of course, maintain their fundamental importance, but now several new issues are getting into the spotlight: support to service provisioning, aiming at the rapid deployment of new services; customer network management; collection of data useful for marketing; and, on the efficiency enhancement side, savings in operations cost that can be achieved by means of automated *operations systems*.

A bit of historical perspective on the underlying technology can help us understand the present state of the art of TMN and perhaps forecast forthcoming developments. About ten years ago, a major breakthrough in the design of network management systems was achieved through the adoption of the *manager-agent object-oriented model*, with its basic idea of defining an *information model* for manageable resources. The consequent advantages in terms of abstraction from unnecessary technological and implementation details, and separation of management applications from basic technology design, were of exceptional importance. The rapid success of the *Simple Network Management Protocol* (SNMP) in implementing this model demonstrated how this much-needed technical development solved real problems

While the SNMP protocol soon reached a diffusion that exceeded general expectations, implementation of *Open Systems Interconnection* (OSI) management-based TMN was seriously slowed down by a number of factors. The availability of standards and the clear advantages of the object-oriented information model did not avoid delays due both to technical and economical factors. Such factors included the inherent cost and complexity of the technology, the lack of application development environments, and the costs involved in substituting the OAM systems in place for the existing technologies with new, standards-based ones. As a result, TMN implementation has been deferred to accompany the deployment of new network technologies.

It is important to realize that, at this point in time, most of the above delaying factors are no longer valid. Field experience with OSI-based network management systems exists, while platforms and development systems have reached a satisfying degree of maturity. New network technologies, designed together with their TMN-based information models, are being introduced everywhere in public networks: *Synchronous Digital Hierarchy* (SDH) transmission systems, wireless communications systems, fiber-rich access networks, and *Asynchronous Transfer Mode* (ATM) switching and access systems. The ease of TMN deployment has been further improved by the inclusion of the *Transmission Control Protocol/ Internet Protocol* (TCP/IP)-based transport options and by the recent advances in the field of distributed platforms. Henceforth, we are probably about to witness a surge in TMN implementations in public networks.

These implementations will be characterized by the two significant technical trends that can be recognized in network management today: *hierarchical distributed implementations*, and *object-oriented computing platforms*.

The traditional, highly centralized approach to network management reflects a scenario where the cost of memory and processing power is still significant and the concentration of management intelligence in a centralized computer allows sharing of an expensive resource. The continuous decrease in costs has made affordable the integration of sophisticated management features in *network elements*, thus achieving better scalability, higher performance, and capacity to support richer functionalities.

For the computing environment, the maturing of distributed object-oriented platforms, and the establishment of *Common Object Request Broker Architecture* (CORBA), permit implementing large-scale, manageable, and evolvable applications within an environment that offers built-in solutions for integrating legacy applications. Hence, the critical dimension of application management is finally introduced into the operations systems architecture. This approach is now widespread and can be the basis for further exciting developments such as the integration of management and service control environments. This integration is advocated, among others, by the *Telecommunications Information Networking Architecture* (TINA) Consortium, which has chosen CORBA specifications as the basis of its *distributed processing environment*.

Although TMN technical perspectives in terms of standards coverage, computing platforms availability, and development systems are, for the time being, very encouraging, we cannot forget the formidable challenge that is currently facing TMN.

Today, new network technologies are being introduced at a very fast rate, due to competitive push and customer demand for better services. But these new technologies cannot be deployed without the related operations systems in place. Therefore, an unprecedented pressure is being put on TMN management systems to match the evolution speed of network technology. This requires an enhanced role of software engineering to dominate processes such as continuous improvement of application development systems, methodologies to integrate existing applications, and implementation of software quality systems. In addition, further progress is strongly needed in the current design of both network technology and related management features.

The dynamics of the environment and the proliferation of new network technologies impact TMN in another critical aspect: the integration of systems. Today, separate management systems are often dedicated to different technologies present at the same time and interoperating in the network. It is possible to conceive, for instance, a network connection whose path is controlled at the same time by different management systems dedicated to the *Wavelength Division Multiplexing* (WDM) layer, the SDH layer, and the ATM layer. Tasks such as alarm correlation, network reconfiguration, and resource allocation become extremely complex when these different systems operate at different layers without a well-structured framework for tackling problems that span all of them. An effective model for federating various operations systems needs to be developed.

In addition to development speed, the *performance of management systems* is becoming more and more an issue as management features are being perceived as an essential part of the service offering. Transport services, operating systems, and distributed computing platforms for management must be considered together with precise performance objectives in mind.

Finally, we must remember that development of the *regulatory framework* in advanced countries calls for openness in networks. An open interface must offer, in addition to transport (and supplementary) services, access to a set of OAM features whose quality is to become an essential part of the service contracts between providers. This requires the ability to guarantee functionalities and performance of each management system at open interfaces, which are or will be opened.

It is clear that the requirements for next-generation TMN systems will be extremely demanding, and it is also clear that, given the necessary engineering effort, present-day technologies have the capability of satisfying them. Still, it is tempting to look at forthcoming technologies to see whether we can expect some new breakthrough that will make us look at management from a fresh perspective. *Transportable computation* and *intelligent agents'* technologies indeed provide new vistas for the development of network management systems. If we recall that the evolution of TMN has been dominated by the tendency toward decoupling management applications from the proprietary implementation of *Network Elements* (NE), it is natural to look with a lot of interest to the development of management *Application Programming Interfaces* (APIs) for the *Java language*. The ability to download application code allowed by this language can add new possibilities to network management practice. For instance, the code for a test can be synthesized dynamically by a centralized operations system taking into account the present state of the network, then wired to the NE and executed locally. Or OAM applications for a particular class of NEs can be developed later by a third party, not necessarily involved in the development of the NE itself.

As a final remark, in reading this book we should be aware of the future importance of TMN and related technologies. If market developments continue at the same pace outlined at the beginning of this introduction, in a few years we will face a situation where the information transport capabilities will become a commodity, probably priced near its ever-decreasing cost. On the other side, the added value will be concentrated on information services and management features, making these the real core of the telecommunications business, the foundations of its profitability, and the drivers of its developments.

Salah Aidarous
NEC America
Thomas Plevyak
Bell Atlantic

Editors' Introduction

*Implement Complexity
to Realize Simplicity*

This reference book is the second in a series on the broad, complex, and critically important topic of network management. The first book, *Telecommunications Network Management into the 21st Century* (IEEE Press, 1994), dealt with techniques, standards, technologies, and applications. Now we extend the presentation of network management technology into the realm of implementation. Each chapter is a component of an orchestrated collection of original chapters written expressly for this book.

Chapter 1, by Lakshmi Raman of ADC Teleommunications, brings detail and clarification together in describing information modeling and its role in network management. These foundation technologies apply across contemporary leading-edge network management methods. Raman states that information modeling "is not specific to network management applications" but to "building distributed applications in general." Successful information modeling, in any application, is based on a "common understanding of the information exchanged." For network management, a distributed application, "two approaches are available, one evolved from the need to manage simple data communications equipment and the other evolved from the need to manage more complex telecommunications equipment." The object-oriented paradigm is introduced with initial focus on the Telecommunications Management Network (TMN) and its Open Systems Interconnection (OSI) and Common Management Information Protocol (CMIP) basis. Later in the chapter, information models for data communications, notably Transmission Control Protocol/Internet Protocol (TCP/IP) and Simple Network Management Protocol (SNMP), are treated. The chapter concludes with future directions in Open Distributed Processes (ODP) and Open Distributed

Management Architecture (ODMA), which "expand OSI management to allow distribution of resources being managed."

In Chapter 2, George Pavlou of University College London, provides a comparative study of key technologies that comprise the management of telecommunications networks. Pavlou starts with the manager-agent model adopted by ISO/ITU-T OSI Management and Internet Management. OSI management addresses the management needs of OSI data networks and telecommunications environments using an event-driven, object-oriented approach that moves management intelligence close to the managed elements. Internet management addresses the management needs of private data networks (LANs/ MANs) and Internet backbones using a connectionless polling-based approach, opting for agent simplicity. The author introduces the ISO/ITU-T Open Distributed Processing (ODP) framework for specifying and building distributed systems, followed by the Object Management Group (OMG) Common Object Request Broker Architecture (CORBA) where server objects are accessed through interfaces on which operations are invoked by client objects in a location-transparent fashion. The chapter provides an excellent tutorial for those technologies and how they are used in distributed telecommunications systems management.

Chapter 3, by Jerry Power and George Pauthner of Alcatel Telecom, addresses a topic of fundamental importance to network management technology implementation; management platforms. The authors state, in general terms, that "A software platform comprises services and behaviors; it provides a common set of functions for the applications that make use of it." Importantly, for application development efforts, "reduced software cost is achieved in concert with increased functional capabilities since this allows the application development team to focus on solving a business problem and not on solving prerequisite infrastructure issues." The chapter establishes a framework for well-designed platform systems and focuses on TMN and Common Management Information Service (CMIS) with its object-oriented commands and reference to objects defined in Generic Definition of Managed Objects (GDMOs). Management Information Base (MIB) construction, Generic User Interface (GUI) services, protocol support, interworking applications, Application Programming Interfaces (APIs), and other platform methods and tools are discussed, as are standards and platform building blocks. Several case studies of commercial management platforms are given as examples of management platforms.

Chapter 4, by Vijay Garg of Lucent Technologies, Bell Labs, begins the journey from network management techniques and standards to implementation in a specific technology, that is, Personal Communications Services (PCS) Networks. The author states that "A PCS network is designed to provide personal, terminal, and service mobility. Personal mobility is the ability for users and services to remain mutually accessible regardless of their locations. Terminal mobility allows a terminal to access services from different locations, while in motion. Service mobility deals with the ability to use vertical features from remote locations, while in motion." The complexity of PCS network management becomes self-evident. The chapter compares SNMP-2 with CMIP; the TIA and Committee T1 Reference Models in North America, which can each be converted into the other; open interfaces; and requirements, goals, and functions for PCS network management. The author concludes with the suggestion that PCS networks be managed with CMIP-based technology.

In Chapter 5, Yechiam Yemini of Columbia University and Geoff Moss of Motorola provide an interesting view on mobile networks and the challenges facing their operations and management. Their coverage encompasses a broad range of current and future technologies ranging from cellular to satellite networks carrying voice and data traffic. Mobile access, and the area of networks access in general, has been creating substantially novel management problems that cannot be resolved through extensions of management technologies being used in the backbone network. The authors review the features intrinsic to mobile networks and the challenges that have resulted from interactions within the physical environment and handling of dynamic changes at the elemental and network layers.

Yemini and Moss discuss how mobility causes events to propagate among components that need dynamic coordination for detecting and updating to a broad range of operational scenarios. With this in mind, they started their approach on how to monitor and control mobile elements, and use this to provide an appropriate network layer management architecture. They also focus on the requirements for applications services management and the quality of service issues from an end-to-end perspective.

As the authors' analysis makes clear, mobile networks are raising important challenges for the telecommunications industry, in managing the access and application services layer. This chapter represents an eye opener on the opportunities enabled by mobile access technologies as well as the challenges in providing management solutions for operating their complex environments.

Chapter 6, by John Brouse of Jones Intercable and Mohamed Beshir of Nortel, establishes the first approach to define the management requirements of CATV networks. In the last few years, CATV networks have been the primary focus of the telecommunications industry to deliver multimedia services due to its penetration within the residential community. The authors start by providing an overview of the variety of existing CATV networks and their characteristics. Then they define the set of processes, practices, and systems used by CATV operators to manage cable networks. A simple network model is established and used to define the fundamental functional building blocks and their data requirements for the management of these networks. The scope of management includes Customer Premise Equipment (CPE) in addition to the headend, outside plant, and drop. The chapter represents a unique reference for management requirements of CATV networks.

In Chapter 7, Mike Ahrens of Bellcore Software Systems measures the astonishing complexity of Telecommunications Management Systems (TMSs) and underlines the role of architectural integrity in helping to tame that complexity. He takes us back to Fred Brooks' *Mythical Man Month* which warned of the complexity of large software systems. TMSs, an example of Brooks' treatise, must "anticipate and respond to business, market, technology, and organizational forces, faster and more effectively than the competition." TMSs must be achitected in what Ahrens refers to as the Service and Network Initiative (S&NI) life cycle. Each S&NI is a complex business undertaking involving market analysis, technology assessment, service definition, network characterization, vendor selection, service requirements, network requirements, and TMS life cycle. Ahrens explains the organizational model of learning as an extension of the individual model of learning and the distinction between know-why (conceptual staff groups) and know-how (operational line personnel). A key for architecture teams is to achieve architectural integrity by understanding how to transfer both conceptual and operational knowledge as the organization designs for complexity.

In Chapter 8, Enrico Bagnasco and Marina Geymonat of CSELT provide a global view of the telecommunications industry, with focus on European requirements for network management. They describe a new emerging model, driven by the techno-economic infrastructure in Europe. The consolidation of the European Union into a single market has introduced new requirements for management of telecommunications networks. Network management standards represent fundamental requirements for introducing advanced telecommunications technologies and services, which drives economic growth in Europe. The authors review major regional network management initiatives, which are directed by the European Community, and explain the key role of standards for achieving interoperability between service providers and countries, in addition to technologies and suppliers.

Implement Complexity to Realize Simplicity

The co-editors strongly believe that the time has come for organizations, worldwide, to implement the complexity of contemporary network management technologies in order to achieve operational simplification and attain their business objectives. This vital aspect, perhaps more than any other, will enable organizations around the world to be more responsive and competitive into the twenty-first Century.

We are confident you will find this book highly informative and useful.

Lakshmi Raman
ADC Telecommunications

Chapter 1

Information Modeling and Its Role in Network Management

1.1. INTRODUCTION

Telecommunications, information, and entertainment network and service providers are experiencing new challenges today more than ever before due to advances in technology combined with growing demand from sophisticated users. In addition, the regulatory environment in many countries is undergoing rapid change. Competition for providing high-quality services is increasing. Competition introduces the need for exchanging information to support successful operation of network elements not only within a jurisdiction but also between jurisdictions (domestic as well as international). Customer network management is also requested by large business customers. Instead of considering network management as an afterthought, the trend is being reversed; product decisions are being made using network management features offered by the suppliers as a differentiator.

Today network management is provided using a variety of data communications protocols ranging from proprietary to de facto standards such as BX.25. Network management information, also referred to as operations messages, are specified in many cases using character strings with delimiters. This acceleration of communication protocols has resulted in islands being created with dependency on one or two suppliers in order to have interoperable interfaces between network elements responsible for providing the services and operations systems or supervisory systems that manage the network elements. This results in either delaying the introduction of new services or deploying them without adequate management in order to meet market demand.

1

These issues are not specific to network management applications. They have been encountered in building distributed applications in general. The fundamental requirement for successful communication is to have a common understanding of the information exchanged regardless of whether it is between two computer systems, between software processes, or between a software process and a database where the information is stored. This need was identified in database design in the early 1970s, and different techniques for information modeling exist in the literature. This chapter describes how information modeling principles have become an integral part of network management, a distributed application.

Two different approaches are available in the industry for developing information models for network management. One approach evolved from the need to manage simple data communications equipments such as bridges and routers, and the other stemmed from the need to manage more complex telecommunications equipment such as switching and transmission nodes. Even though the actual technique and details differ with the two approaches, the fundamental principle remains the same, namely, model the information across a communication interface so that both the sender and receiver of the information interpret it in the same way. Other information modeling efforts are in progress for development of portable software in distributed applications.

This chapter provides an introduction to information modeling, forming the foundation for an interoperable multisupplier network management solution. Even though several techniques are available for information modeling, the focus for this chapter is on using the object-oriented principles. These principles, which were initially found to provide a powerful foundation to develop reusable software, have been adopted to varying degrees in designing and building open distributed processing applications.

Sections 1.3 through 1.5 of this chapter discuss the need for information modeling to provide an interoperable management interface between the managed and managing systems. An introduction to different information modeling methods is provided in sections 1.6 through 1.8. Components of specific object-oriented information modeling principles used in support of telecommunications network management are discussed in sections 1.9 through 1.18. These components allow modeling complex telecommunications resources. The principles used for modeling data communications resources are simpler and are not discussed here in detail except to note some of the differences. Examples of information models available to support network management of telecommunications and data communications equipments are provided in sections 1.19 through 1.25. The goal of network management is to facilitate deployment of interoperable network equipments. Sections 1.26 through 1.29 discuss mechanisms available for suppliers to specify conformance of their products to requirements and for service providers to determine interoperability issues prior to testing and deployment. Advances in software development, specifically distributed processing, is influencing the future direction of network management specifications, specifically the Telecommunications Management Network (TMN). Since network management products are expected to embrace distributed processing concepts, probable future directions are discussed in sections 1.30 through 1.34. Conclusions are provided in section 1.35.

1.2. INFORMATION MODELING MADE EASY

Before discussing the various object-oriented concepts used to develop an information model, a simple approach to defining an object as part of an information model
is presented in this subsection. An information model for an application may be
composed of one or more objects to meet the different requirements. Because this
chapter focuses on the role of information modeling in network management, examples are taken for that distributed application. The need for representing a resource
in the information model must first be established. In other words, the object must
meet some requirement for the application.

In developing the information model for network management applications,
only the information that is relevant to management is modeled. For illustrative
purposes, consider that it is a requirement to manage a hypothetical line card in a
switch. The line card is a physical resource that exists to support call processing
activities. In developing the information model for NM, only the information about
the line card relevant to management is modeled.

Assume that the properties of the line card to be managed (either for the
purpose of monitoring or control) are the following: the line card is produced by
some supplier; it has a serial number, an equipment type specifying the type of line
card; and a state indicating whether or not it is active; and, if the line card is active,
then it is assigned to some telephone number. Information such as alarm status and
equipment type will be required when the line card has a fault and needs to be
replaced as part of fault management application. On the other hand, the telephone
number property is required to provision a subscriber. The information model for
representing these static properties may consist of the template illustrated in Table
1.1. The table identifies how these different properties are represented so that a
common understanding exists between the managed switch where the line card is
present and the managing system. The representation identifies the syntax of the
information at communicating interface, and the properties describe the semantics of
the managed information.

If the line card fails, it may emit an alarm message that will be sent to a Network
Management System (NMS), and the NMS may request the following actions:
activate/deactivate the card and replace the faulty card with another card of the

TABLE 1.1   TEMPLATE FOR A LINE CARD

Properties	Representation
supplier name	character string
serial number	character string
equipment type	character string
card active state	yes/no
alarm status	critical (0)/clear (1)
assigned phone number	numeric string
:	:

same type. In addition to the static properties included in the table, ability to emit an alarm notification will also be included as part of the information model for the line card. These properties of the line card can be applied to every line card, regardless of how it is physically implemented and in which switch it is present. The above template represents a class of objects called the line card.

Every actual line card in the switch is represented by one instantiation of the template described in Table 1.1. Table 1.2 illustrates such an instantiation.

The properties with specific values shown in Table 1.2 represent a line card regardless of the actual implementation of the card and the supplier of the card. The power of information modeling is to bring together in a representation characteristics of the interface an object supports, regardless of the actual implementation. A collection of instances belonging to different object classes form a repository. In the case of network management, this repository is referred to as the Management Information Base (MIB). The collection of objects that represent the entries in a directory is known as the Directory Information Base (DIB).

TABLE 1.2   A SPECIFIC LINE CARD OBJECT

supplier name	ADC
serial number	cu126781
equipment type	ISDN Channel Unit card
card active state	yes
alarm status	clear (1)
assigned phone number	946 2090
:	:

1.3. COMMUNICATING MANAGEMENT INFORMATION

Network elements are managed today using different methods varying from proprietary to open interfaces, as pointed out in the previous section. Because this chapter is concerned with operations information being transferred at the application level,[1] methods such as E-Telemetry are not discussed. Two methods are currently in vogue to manage the network elements that pertain to sending application-level information. The first method is a message-based paradigm, and the

[1] Application level is used in the context of the OSI Reference model. The information is exchanged between application processes in the OS (Operations System) and NE (Network Element).

second is an object-oriented paradigm. The object-based approach is used for managing both telecommunications and data communications resources. However, the actual details of what the object represents vary considerably in both cases largely because of the differences in the interface definition.

1.4. MESSAGE-BASED PARADIGM

Several systems are deployed today in the service provider's network that use the message-based approach. The information exchanged is specified in terms of messages. Languages used to specify the messages use a human friendly format. Usually, character strings are used to define the exchanged information. As an example, in North America, message sets are available as open generic specifications using a language called Transaction Language 1 (TL1) developed by Bellcore. The message sets are defined in Bellcore Generic Requirements for applications such as alarm and performance monitoring, testing, and provisioning. The messages are specified using either position-based values for management information separated by delimiters or a tag value scheme.

In either scheme the messages specify the type of operation or notification along with one or more entities being managed.[2] Taking TL1 as the language, we provide two examples of the messages between OSs and either NEs or Mediation Devices (MDs).[3]

```
^^^RDBKNJ35672^85-04-10^05:46:20
*C^101^REPT^ALM^T1
^^^''101:CR,T,SA,FEND,,261,259
(^ is used to indicate space)

ENT-LI:267943:143: :OE=003151026,MC=1P,CHT=FL,PIC=123
```

The first example specifies the content of an alarm report on the entity referenced as T1 101 reported by an NE to an OS. If an alarm is to be reported on a different entity such as a circuit pack, a different message will have to be specified. In other words, for the same type of notification or operations request,[4] because the managed entity is different, a new message will have to be created. When developing the software, except for the parsing routines, it may not be possible to reuse code designed to support one message with another. Powerful software engineering con-

---

[2] Bellcore requirements define messages for most of the applications except provisioning (also referred to as memory administration). In the latter case, a nonnormalized data model combined with verbs corresponding to database operations define the messages.

[3] Mediation device is sometimes referred to as supervisory system.

[4] The terms *notifications* and *operations* are used in order to conform to the terminology used in TMN standards. In TL1 these are usually called commands and autonomus notifications, respectively.

cepts such as abstraction and reuse are inherently not available with this paradigm. Specification of the complete message structure, however, makes it easy for a reader to understand the exact information exchanged without requiring knowledge in the intricate details of the protocol and for programmers to develop the software specification.

In the second example using the relational data model, an OS sends information on a subscriber to a switching system. The verbs used with the management information in the data model correspond to the basic database operations.[5] Defining the data model and using a defined set of verbs are appropriate for memory administration, given the large number of features available with the Integrated Services Digital Network (ISDN) and Advanced Intelligent Network (AIN). The database operations themselves are not different. The information entered in a switch varies depending on the services selected by the subscriber.

The message-based paradigm in general was developed to provide the same message definition for user-to-machine and machine-to-machine interfaces. The language chosen in many cases is derivative of the ITU Recommendation Z.300, known as Man Machine Language (MML). Another advantage of this approach is the ability to use a simple protocol analyzer on the communications link to verify the content of the exchanged information. Using a user friendly message specification viewed from the user perspective has disadvantages from the machine-to-machine point of view. The languages employed are constrained to use human readable character sets. This restricts the data types for message specifications. Other data types such as integer and boolean that are more applicable for machine operations are not used.

Even though systems using the message-based paradigm are deployed extensively, this approach lacks rigor in specification. As an example, in referencing an entity being subject to management, it is necessary to provide an unambiguous identification. In the TL1 approach, the specification assigns n number of characters to be used by the supplier. The format for the identification is influenced largely by the architecture of the supplier's product. Even across multiple network management applications, different formats have been recommended.

1.5. OBJECT-ORIENTED PARADIGM

In the telecommunications environment, the equivalent of the "messages" is specified using an object-oriented paradigm. Several information models using this paradigm have been developed as part of standards work on the Telecommunications Management Network (TMN). With differing details and complexity, object-based approaches are defined for managing Internet data communications network resources as well as for building software for distributed applications. The following subsections describe the approaches used in telecommunications and data communications network management.

---

[5] The database operations are referred to as CRUD to denote create, read, update, and delete.

1.5.1. System Management Architecture

The architecture for exchanging management information for interactive applications is shown in Figure 1.1.

In Figure 1.1, let us assume that an OS is managing a network element; this implies that the OS is the managing system and the NE is the managed system. In contrast to the message-based paradigm, the object-oriented approach specifies a set of remote operations.[6] These remote operations may be performed on different resources depending on the technology, services, and architecture being managed. The managing system issues operations requests on resources and receives notifications corresponding to the various events. In other words, taking the example used in TL1, the structure of the message for a notification is not redefined if the notification is an alarm from a termination point or a circuit pack. Similarly, the structure of the message to create a subscriber is not different from that of a log. Depending on the resource being managed, the properties to be sent with the create request will vary. Instead of defining new messages, the resources are modeled as managed objects with specific characteristics.

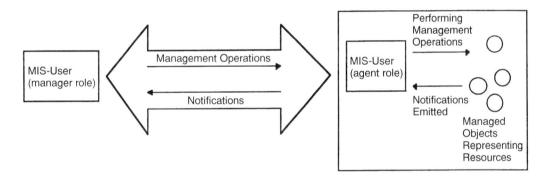

Figure 1.1 System Management Architecture.

1.5.2. Message Structure for Telecommunications NM

Common Management Information Protocol (CMIP), specified to provide the services defined in Common Management Information Service (CMIS),[7] is used for external communication. The various services offered by CMIS include the database

---

[6] The term *remote operations* is used here to distinguish it from commands where, for example, an OS requests the agent to perform a specific operation such as creation of a subscriber record. Both commands and notifications are to be considered as remote operations.

[7] Common Management Information Service Element (CMISE) is an OSI application service element consisting of a service definition known as CMIS and a protocol specification called CMIP.

operations to create, retrieve, delete, and modify data, along with resource-specific event reports and actions. Because the database operations are generic, depending on the resource, additional definitions (event and action types) will be required. A high-level structure for the "message" is shown in Figure 1.2.

Every message has a sequence number[8] and a value or code to identify the remote operation. A specific set of operation values are defined as part of CMIP, and these are reused across multiple management applications functions. For example, the same operation value and structure is used in the message to replace the value of window size of a protocol entity and the state of a cross connection. The resource and the property being modified vary with each message. The resources and their properties that can be managed are specified in terms of object-oriented information models. In other words, the message structure leads naturally to an object-oriented paradigm.

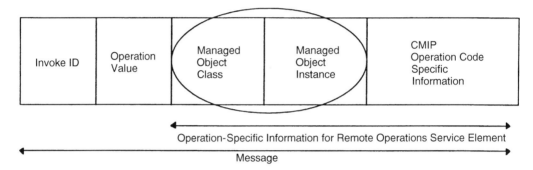

Figure 1.2 CMIP Protocol Data Unit (Message Using CMIP).

1.5.3. Message Structure for Data Communications NM

The system management architecture described in Figure 1.1 is suitable to describe the network management of data communications network, specifically the Internet. The concept of operations to be performed on objects, known as MIB variables, is the same, even though the types of operations are not identical. The ability to report notification is equivalent to trap messages indicating something has happened. Again the details of the types of notifications and the information exchanged with it are considerably different. The structure for the messages is shown in Figure 1.3.

Request ID	Object Name	Object Syntax

Figure 1.3 SNMP Protocol Data Unit (Message Using SNMP).

---

[8] The concept of sequence number to allow correlation between the request and response message is also present in the message paradigm such as TL1.

The figure is simplified in that only one reference to object name and syntax are shown. The repetitive nature of these two parameters is not illustrated in the figure. In addition, the type of operation is implicit by the tag of the message instead of an explicit operation value parameter shown in Figure 1.2.

Similar to CMIP, the number of operations is finite, and various requirements of network management are achieved by modeling the resources as objects. (The combination of object name and syntax is referred to as varbind.)

It should be noted that modeling application-specific information using an object-oriented approach does not impose requirements on software development. In other words, there is no requirement to implement using object-based languages. However, there are benefits to be gained in software development by object-based languages because the same principles are used.

Even though the concept of object-oriented modeling of information is discussed here in the context of message exchange, information modeling is not specific to external communication. While it provides a rigorous formalism to define the properties of managed resources for the purpose of network management communication, efforts are underway in ITU as well as other organizations such as the Object Management Group (OMG), to use an object-oriented approach for information modeling in general.

1.6. FOUNDATIONS OF INFORMATION MODELING

As mentioned earlier, defining information models, also known as schema, is an activity that is not specific to network management. Data models using entity relationship concepts have been in use for a few decades. Entity relationship models define various entities and how they are related to each other. This section provides a brief introduction to two modeling approaches. Instead of discussing how to develop information models in general, emphasis is on how to define the schema within two network management contexts (datacom and telecom). Because the principles used in telecommunications information model are more complex, these are discussed below. Differences with the modeling principles used with data communications network management are noted.

A technical report published by ANSI T1 provides guidelines for specifying an information model. Several concepts that aid in developing an information model (specifically an object-oriented design) are introduced in the following sections. How these concepts are used in developing an information model depends on several criteria. Some questions to consider are:

- Is the model easy to understand?
- Is it implementable?
- Does it meet the requirements set forth at the start of the modeling process?
- Is it extendible and does it meet multiple supplier solutions to a technology?

The ultimate test is whether the model can be implemented efficiently to solve the specific application, for example, in the context of network management. It is possible to define a model that meets all requirements and is very elegant from the specification point of view. However, implementing this model may impose heavy processing and memory requirements. This will result in degrading the performance of the network being managed, thereby affecting the quality of services provided. Such a model will not be considered an acceptable specification. In developing information models for use with SNMP, simplicity was a main goal. *The Simple Book* by M. T. Rose discusses criteria for including an object definition in the schema. Even if the same criteria are not suitable in all applications, setting such guidelines can provide the balance between architectural purity and practical implementation.

In setting guidelines, one encounters requirements that may be complementary or conflicting. Understanding the tradeoffs and optimizing to meet a certain criterion will be necessary. In other words, it is fair to state that modeling is an art and there are no right or wrong answers.

1.7. E-R APPROACH

Early attempts at information modeling were done in database applications. As such, relationships between the components of the system were defined. Several books have been written on the topic of information modeling concepts (e.g., Matt Flavin, *Fundamental Concepts of Information Modeling*), and emphasis has been on identifying the relationships between the components of the data.

In the entity relationship approach, as the name suggests, the model defines various business entities and relationships between them. This approach to facilitate database design was first introduced by Chen (P. P. Chen, "The Entity Relationship Model—Toward a Unified View of Data," *ACM Transaction Database Systems*, Vol. 1, No. 1, 1976) and extended by Codd later (E. F. Codd, "Extending the Database Relationship Model to Capture More Meaning," *ACM Transactions on Database Systems*, Vol. 4, No. 4, 1979). Even though this approach has encountered criticism, the fundamental reason for its popularity is its simplicity. Identification of what constitutes an entity and what relationships should be modeled is somewhat subjective.

In addition to defining the entities and their relationship, these models are accompanied by diagrams describing the relationships between the entities along with the type of the relationship. Even though the object-oriented paradigm used in TMN information models has followed a different approach, standards have used an E-R like diagram to make the model human friendly. In other words, the notation used to define the models facilitates machine parsing rather than making it easy for a reader who is attempting to get an overview of the various objects and the relationships between them.

1.8. OBJECT-ORIENTED DESIGN

Unlike the E-R modeling approach, object-oriented design principles, simply stated, provide an abstraction that encapsulates data as an object. In terms of network management, the resources managed are specified as managed objects. In most cases, the resources exist to provide telecommunications service(s). The example of line card template in Table 1.2 is a managed object. Managed objects provide an abstraction for the properties of the resource that are manageable. In other words, details not relevant to management are not modeled.

The advantages of the object-oriented approach to model information are encapsulation, modularity, extensibility, and reuse. With the concept of encapsulation, the object is responsible for assuring integrity when it receives requests (messages) to perform operations. In other words, internal details of how the operation is performed by the resource is not visible at the boundary of the object. The abstraction of the resource defined as an object specifies the properties that reflect the result of executing the message.

The object-oriented approach naturally lends itself to modular specification. What aspects of management information should be considered as an object depends on the level of modularity desired. This can be illustrated by using the following examples. Let us consider the case where performance monitoring parameters are to be collected from a Synchronous Digital Hierarchy (SDH) line termination. An object may be defined to represent all the information, including the performance monitoring data pertinent to that resource. A more modular specification will separate the fundamental properties such as state from Performance Monitoring (PM) parameters. By including all the appropriate PM parameters within an object, the specification becomes modular, which automatically leads to flexible and extensible specification. New PM parameters may be introduced without affecting the definition of the termination point. Another example is provisioning a customer's subscription data for various services. Here again, modeling a subscriber with parameters corresponding to all the requested services will not provide a modular specification. Instead, modeling the parameters of a service as a separate object related to the subscriber results in a more modular specification. The flexibility is not just at the specification level, but it can also be reflected in software. However, flexibility is not without additional cost because the number of objects and relationships to be maintained increases. By appropriate design, modularity facilitates reuse both at the specification level and in software. Reusability at the specification level and in software is discussed in this and the following sections.

The paradigm used for defining SNMP information models does not use various object-oriented principles such as inheritance and encapsulation (identified above). The objects are simple atomic data elements. A complex structure such as a routing table is modeled as a table with various columns.

Information models developed by OMG/CORBA for the purpose of object-oriented software development use many of the principles mentioned above. Even though there are differences in the details of the modeling principles and information models, a common thread among these different applications is the need for an information model. The models provide for interface definitions and facilitate exchanging information in an unambiguous manner.

Development of complex distributed systems necessitates system engineering requirements that include well-specified interfaces. The power of information modeling makes it a strong component of this upfront engineering and can potentially reduce development costs.

1.9. INFORMATION MODELING PRINCIPLES

Several variations of object-oriented design principles are being used for both information model specification as well as in software development. In this section, only the object-oriented design principles defined in ITU Recommendation X.720 | ISO/IEC 10165-1 are discussed. The term *Structure of Management Information* (SMI) is used to refer to the various concepts, as well as the representation techniques used to specify the information model. These principles form the basis for the information models that are available today as part of TMN standards. The subsections below describe various concepts that are used to design an information model. However, it is worth repeating that modeling is an art. As seen in the following discussion, there is more than one way to model the functions of a resource for management. Considerations such as optimizing data transfer for external communication, ease of retrieval and manipulation of data, granularity level for the information, and possible implications for implementations (memory size, processor power, speed of processing) will drive the decision between multiple choices.

1.10. MANAGED OBJECT CLASS DEFINITION

A managed object, as mentioned earlier, represents the management aspects of a resource. A resource may be physical (e.g., circuit pack) or logical (cross-connection map). Taking the example of the circuit pack, several managed objects[9] may be present in a network element to represent the various cards. A managed object class called circuitPack is used to define the properties that are common across different cards. Characteristics such as operational state identifying whether or not the circuit pack is working, type of alarm to inform users that the circuit pack has failed, and behavior stating that they are replaceable plug in units are applicable to all circuit packs regardless of the supplier and the actual function supported (power supply, line card or processor, etc.).

Generalizing the above example, we can state that a managed object class defines a schema or a template with properties shared by managed objects that are instances of this class. Figure 1.4 shows the difference between the managed object class definition and a managed object (instance) for circuit pack.

---

[9] The term *managed object* is sometimes referred to as a managed object instance.

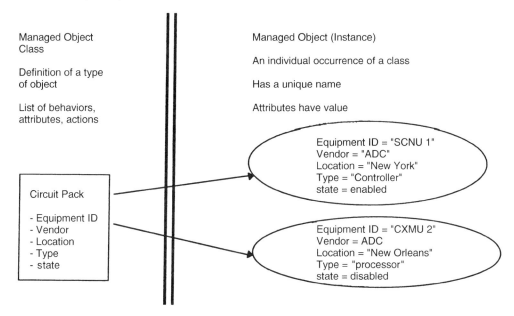

Figure 1.4 Example of Managed Object Class versus Instance.

A template for a circuit pack managed object class includes, in addition to the behavior, the following attributes: circuit pack ID to support unambiguous identification of a specific circuit pack, operational state to indicate whether an instance is working or has failed, and the type of circuit pack; and the following notifications: an alarm to inform the OS that the circuit pack has failed and a state change notification to indicate that the state of the circuit pack has changed to the value "disabled." Let us assume that an instance of a circuit pack is created to represent a power supply in a network element. The instance at any time contains values for the attributes and emits notifications whenever the events resulting in the notifications occur. In this example, all the properties defined in the template are expected to be present in an instance. Modeling of functions that are optional in a resource is discussed below.

1.10.1. Package Definition

The characteristics of a managed object class are defined in terms of behavior, attributes, notifications, and operations. In order to provide variations among instances, the concept of "package" has been introduced. A package is a collection of characteristics (behavior, attributes, operations, and notifications[10]) as shown in

[10] The semantics of attribute, operations, and notifications will be discussed later. For example, state representing whether the resource is active or standby may be modeled as an attribute, requesting a loop-back test may be modeled as an operation, and reporting a circuit pack failure, as a notification.

Figure 1.5. It is not required that every package include all the characteristics. However, if an instance includes a package, all the properties defined for that package must be present. In other words, a package is atomic, and further breakdown is not permitted.[11]

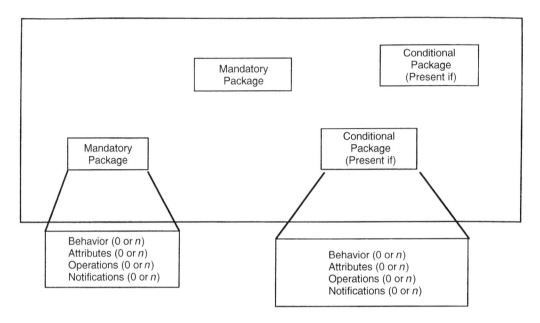

Figure. 1.5 Managed Object Class Definition.

Two categories of packages may be used in defining a managed object class. A *mandatory package*, as the name suggests, includes all properties that are present in every instance of that class. The second category is referred to as a *conditional package*. The properties defined for this category are included in a managed object during its creation if the condition for their presence evaluates to true. This concept of conditional packages has been used also to account for optionality where the condition is a user's option. As an example, consider an object class that represents an NE. Alias names may be assigned for the NE by the supplier to provide a human friendly name. The attribute for the alias name is modeled as part of a conditional package where the condition statement implies this is optional.

The concept of packages has been included in the information modeling principles in order to identify the collection of properties that may or may not be present. It should be noted that packages are defined only to aid in specification. A package does not exist outside of a managed object; properties included in a package are present only as part of a managed object. Properties belonging to packages can be

---

[11] In order to claim conformance to a definition, all properties of a package are required to be implemented. However, when phasing implementations, agreements may be made to include only a subset between suppliers of the managing and managed systems.

included in a managed object only at creation time. In other words, once a managed object is created, the properties of the object become part of the object and are available at the object boundary. They are not referenced or separated in terms of package boundaries. As a corollary, properties belonging to a package cannot be added after the creation of the managed object. If this is required, the original managed object should be deleted and a new one with the additional properties must be created. Similarly, during the lifetime of a managed object, existing properties cannot be deleted either individually or in terms of packages. As with addition, the deletion of properties also requires the deletion of the managed object followed by a creation without these properties.

A package is included only once within a managed object. In other words, multiple copies of a package cannot be present within a managed object. As an example, consider the conditional package mentioned earlier for alias name. Let us assume that it contains an attribute for a user friendly name. When the managed object class with this package is instantiated, the attribute is included only once. This is true even if there are two packages with the same attribute. Even though the specification gives the impression that there are two copies of the same attribute, when the object is instantiated, only one copy is present. This is consistent with the earlier discussion that knowledge of package boundaries is present only at specification time; the boundary is not maintained once the object is instantiated. When two packages have the same property such as an attribute, it is the responsibility of the managed object class designer to ensure that there are no contradictions.

The next subsections define the various components that can be included in a package definition.

1.10.2. Behavior

Behavior definitions are used to describe, for example, semantics and integrity constraints to be satisfied when the components of the package are included in a managed object. Even though specific behavior may be associated with each attribute or notification, when present at the package level, behavior definition is the glue to bring all the properties together. The circuit pack object class specified earlier, includes an attribute for the state and an alarm notification when there is a failure. The behavior describing that, when a critical failure occurs in the circuit pack, the value of the state attribute should be changed to the value "disabled" correlates the effect of an event to the modifications in the value of an attribute. Integrity constraints may be specified in terms of pre-, post-, and invariant conditions. Let us consider the case where the values of two attributes such as window size and packet size for a protocol entity are constrained by a discrete set of values for the ratio. The behavior definition is used to describe an invariant condition that the changes to the attribute value should not violate the integrity constraint on the ratio.

In addition to providing constraints among the properties defined within a package, behavior definitions may also describe how characteristics within a package may be influenced by the presence or absence of a conditional package. As an illustration, consider a managed object class definition that represents a log. Let us assume that an attribute called availability status is defined as part of the mandatory package. This attribute in general may have several values to indicate multiple status information. The value "log full" is applicable for all instances of the log object. Suppose there is a conditional package that permits logging at scheduled intervals (during off hours 5 P.M. to 8 A.M. when no one is available to monitor alarm displays). Then behavior will be added to indicate that "scheduled off" is another valid value if the instance includes the conditional package for scheduling. As a result, when scheduling of the log is permitted, the availability status may assume one or both of the values "log full" and "scheduled off."

The examples provided here pertain to behavior corresponding to characteristics within a package or across packages within a managed object class definition. Another context in which behavior definitions are applicable is to describe the effects on a relationship among objects of the same class or different classes. For example, suppose an equipment holder managed object class represents a slot where a circuit pack can be plugged in. When the circuit pack is absent in the slot, the value of the attribute holder status will be empty. However, when a circuit pack is inserted, resulting in either creating a circuit pack managed object or updating the state of an existing managed object (for example, an existing circuit pack was pulled out and replaced), then the holder status must be updated to a value "occupied." Thus, behavior within a package definition is used to describe effects that are applicable to the managed object itself.

1.10.3. Attributes

A package includes zero or more attributes that reflect static characteristics of the object. In other words, properties described as attributes are distinguished from dynamic or active aspects such as notifications and ability to perform operations. An attribute is defined in terms of an identifier and has a value in a managed object.

The value of the attribute is determined by the syntax defined for external communication. The pair of items (attribute ID and attribute value) is sometimes referred to as Attribute Value Assertion, or AVA. That is, an attribute is asserted to possess a certain value. In addition to defining the syntax for external communication, attribute definitions may also include the types of checking appropriate for the values. For example, if the syntax of the attribute is defined to be an integer or real, then the value can be checked to be greater, equal, or less than a specific value. If the syntax is a complex structure consisting of multiple elements, then simple matching criteria will not be applicable. In some cases, special rules can be specified using behavior statements on how the values are to be checked.

Even though attributes are used to define properties such as state and circuit pack type, it is possible to indirectly invoke an operation by setting the value of an attribute to a specific value. A managed object class defined to represent processors may include an attribute called "restart" with syntax being Boolian. The behavior of this attribute will indicate that setting this value to true requires that the processor be restarted. That is, the indirect effect of changing the value of the attribute is to start a process.

Often, it is possible to specify the syntax, matching rules, and behavior specific to an attribute without taking into account the package (or indirectly the managed object class) where the attribute is included. The definition of the attribute by itself addresses to a large extent the syntax.[12] The semantics of what the attribute represents within the context of the package (actually, the managed object class representing the resource) is determined by the characteristics of the resource.

Attribute-oriented operations[13] are specified to reflect whether in the context of a specific managed object, the value of an attribute is read only versus modifications are permitted. It is possible that the same attribute may be allowed to be modified within one managed object class while this operation is not permitted in another object class. As an example, consider the counters for performance monitoring (PM) parameters. Let us assume that two classes of objects are defined to contain the PM parameters. One class is used to contain the values corresponding to the current collection interval, and another class contains values collected in the previous interval (historical information). Attributes for PM parameters, such as coding violation, will be defined to be readable as well as resettable to allow zeroing them at any time during the collection interval. However, when the same PM parameters become attributes of the class representing the historical data, then the only allowed operation is read. Any modification of historical information will be disallowed. It is the responsibility of the object to ensure that the integrity constraints defined using the pre-, post- and invariant conditions in behavior are not violated as a result of performing the requested modification.

Depending on the syntax, the value may be a single value (corresponding to, say, an integer syntax) or an unordered collection[14] of values. The latter is referred to as a "set valued" attribute. Additional operations such as adding and removing values to the set may be included with this type of attribute. The set of values is treated as a mathematical set. Adding an existing value, while not an error, will not include a second copy. Similarly, removing a nonexisting value is not considered an error.

Even though the value of an attribute is determined by its syntax, within the context of a managed object class, restrictions may be imposed on the specific values or range of values allowed for that resource. Two types of value restrictions may be specified. These are not mutually exclusive; either or both may be present in any managed object class definition. The set of permitted values is used to specify all the values that a resource may support. The set of required values specifies the minimum an instance of this class should support. For example, assume that an object class is defined to represent a modem with an attribute called "speed." Let us suppose that the syntax is real. A standard set of values such as 2.4, 4.8, 9.6, 19.2, 56, and 64 Kb are permitted for any instance of the modem class. Any other real value for the speed

---

[12] Behavior definitions may be used to specify semantics.

[13] Use of the term *attributed-oriented operations* should not be taken to mean that the operation requests are issued directly to the attributes. These are operations performed by the managed object affecting the values of attributes they contain.

[14] Multiple values that are to be considered in a specific order (for example, sequence of integers increasing in value) are treated as a single value.

will not be allowed. However, all instances may be required to support 9.6 Kb. This implies that in order to be conformant to this specification, all managed objects implemented to this definition must be capable of supporting 9.6 Kb. This example also points out that the set of required values must be either a subset of or the same set as the permitted values.

Another type of restriction that can be specified for the value of the attribute within the context of a managed object class is identifying a default value or an initial value. The difference between them is as follows: a default value implies that when the object is created and the value of this attribute is not provided, then the default value is used. This situation occurs irrespective of whether the attribute is part of a mandatory package or present in the conditional package that will be included (because the result of the condition evaluates to true). When an attribute has an initial value, this implies that during the creation of the managed object that attribute will not be permitted to have any other value. If a request to create supplies a value other than the initial value, the object will not be created. However, when a value is not supplied, the initial value is used and the behavior is similar to the default value.

In addition to providing a default value to use in the absence of one being specified in the request, default value can be used to reset the value at any time during the existence of the object (assuming no other integrity violations are encountered). Continuing the example of performance parameters, it is essential that counters representing the PM parameters are set to an initial value of zero before starting data collection. However, it is also true that the default should be zero so that, at any time, the OS may request a reset of the counters. The combination of both default and initial values in the specification will be required to meet the two requirements.

1.10.4. Attribute Groups

Attribute group is a handle to refer to a collection of attributes. It is a shorthand form to request a group of attributes. An attribute group has an identifier similar to an attribute. However, unlike an attribute, it has no value of its own. Restrictions as to what attributes may be included in the attribute group depend on the type of attribute group and are discussed below. The handle may be used with only two operations—read and set the value of the individual group elements to default values. In order to support the latter function of setting to default, the individual attributes included in the attribute group definition should have default values associated with them in the context of the specific managed object. With the read operation, instead of including the identifiers of all the attributes, referencing the identifier corresponding to the group will return the values of all the constituents. There is no limit on how many attribute groups may be included within a package.

Two categories of attribute groups may be defined: fixed and extensible. The fixed group, as the name implies, is a definition (the set of attributes that are included in the group) that cannot be modified later. If new attributes have to be added, then another attribute group must be defined. In order to include an attribute as a member of a fixed group, that attribute must be present in the same package as the attribute group.

Extensible group, on the other hand, refers to an attribute group where new attributes may be added after the original definition. These new attributes (because the original managed object class was extended to include additional properties) should either be present in the conditional package where the attribute group is included or be part of the mandatory package.

Let us now look at how this may be used in a model. Consider the case of performance parameters corresponding to DS1/E1 line included in a managed object class. All the parameters have a default value of zero. A fixed attribute group can be used to optimize data transferred in a request. Instead of listing all the PM parameters, a handle using the identifier of the attribute group can be provided to read all the parameters. Similarly, all the parameters can be set by using the set to default operation on the attribute group.

An example of use of extensible attribute group in the literature is for state attributes. Even though several state and status types are defined, not all are applicable in the context of a resource. Moreover, additional state/status types are expected to be defined that may be specific to a resource. In this case, a simple form of extensible attribute group is one with no attributes. Behavior is defined as follows: "components of the group are determined by the state and status attributes for the class containing this group." In other words, the elements in the group are dynamic. If this attribute group is included within a managed object class definition, then referencing this group for a read request will include, in the response, all state/status attributes present within the object class. If this group is included in the mandatory package, then the elements in the group may vary between different instances depending on whether state/status attributes are present in conditional packages that may or may not be included when the managed object is created.

1.10.5. Notifications

The previous subsections addressed the matter of how to model characteristics that are appropriate for database operations.[15] Let us now discuss characteristics that are modeled using the concept of notifications. These are triggered by either internal or external events.

Consider the case of internal events. Such events may occur for several reasons. Addressing the model of a circuit pack to represent a power unit, we see that failure of the unit or the board itself will result in generating an event of type equipment alarm. As a result of this alarm, the state attribute of the circuit pack will change to disabled. Suppose this power unit has a backup; then an automatic protection switch to this backup unit may occur. The status of the circuit pack managed object representing this backup unit will change from "standby" to "active." All three events, equipment alarm and two-state change notifications (state and status), are generated

---

[15] The phrase "static" has been sometimes used to describe attributes. An attribute, such as a PM counter, is dynamic in that the value is changed as a result of, let us say, detecting an errored second. Another way to think of attributes is information that is always present once included while creating the managed object.

as a result of internal events power unit failure and protection switch. The definition of the managed object class will include these notifications (as part of the mandatory or conditional package).

Examples of notifications arising as a result of external events are the following. Suppose two operations systems (OSs) are used in managing an NE. Assume that one of the systems is concerned with configuring the NE and the other with obtaining alarm information. The OS configuring the NE may send a request to create a circuit pack object when a new line card is plugged in a slot. When a circuit pack object is created in response to this request, an object creation notification will be generated by the newly created circuit pack to announce itself. This event may then be forwarded to the OS performing the alarm monitoring function so that it can understand alarms from this newly created object. Another example commonly used is the request to modify the value of one or more attributes. This is again an external event that gives rise to a notification referred to as attribute value change.

When the managed object class (through the package definition) includes a notification, behavior is used to explain the circumstances for generating the event. In addition to referencing the type of notification, it is also necessary to include the syntax for the information associated with the notification in order to communicate the event to an external system. For example, the parameters for an equipment alarm may include severity, probable cause, diagnostic information, and whether the resource has been backed up. Specific data types must be defined, similar to the syntax mentioned for attributes, in order to communicate details of the event to an external system. When specifying syntax of the notification, to promote reuse of the definition without restricting the ability to include additional information, it is recommended that extension capabilities be included in the syntax. The syntax of the extension itself will have to be determined once required extensions are identified.

In TMN modeling, it should be noted that occurrence of an event within a resource resulting in a notification must not be taken to imply that this is always communicated via the protocol to an external system. The next section defines a mechanism available in standards, and it is possible for the managed or managing system to configure the criteria under which a notification is communicated to a specific system.

1.10.6. Actions

This concept is used to model operations that the resource is capable of performing at the request of a managed system. These operations are distinguished from the attribute-oriented operations mentioned earlier. The latter refer to requests to read or modify attributes; even though the request references the object, the requested operation is on the attribute(s) only. Actions on the other contain requests to perform operation on the object as a whole. Associated with an action type is information required to perform that action successfully. The syntax of the information will have to be specified to communicate the request externally. One or more replies may be present for an action request. Syntax specification will be required to explain how to interpret the response containing the result of performing the action request.

An action definition is usually accompanied by a description of the process for performing the operation. In discussing the various kinds of modifications on the attribute, it is noted that when attribute values are modified there may be indirect effects, as in the case of replacing the value of the restart attribute. That is, it should not be misconstrued that action is always needed when describing a process. Defining an action to describe a process does not provide any additional feature beyond what is available if modeled as a result of the modification of an attribute.

Examples of where action instead of set[16] is more appropriate are as follows: information issued in the request is not required to be an attribute of the object,[17] operation requires coordination of activities across multiple objects that may or may not be of the same class, information in the reply may not correspond to attribute values. The request to perform a test is modeled using an action. This is more appropriate because often initialization information sent in the request is required to begin the test. In order to perform a connectivity test on a termination point, the remote end will be provided. This is not required to be an attribute for the lifetime of the managed object but information required to start the test. Similarly, the result of the test "pass," "fail" is not appropriate to be modeled as an attribute. When using action to model the test request and response, it is also possible to send multiple responses indicating progression of the test. All these responses can be related to one request. X.722|ISO 10165-4 describes guidelines for using action versus set. As stated, they are only guidelines; it is still the judgment of a modeler(s) or a compromise between different members in a standards environment that influences the decision.

As with notifications, it may be appropriate to include extension capabilities with the syntax definition for the request and response to an action.

1.11. SYSTEMS MANAGEMENT OPERATIONS

The sections on attributes, attribute groups, and actions discussed two kinds of operations—attribute-oriented, where the database operations such as read and modify the values of an attribute are applicable, and object-oriented, where action requests are addressed to the managed object as a whole. These operations are sometimes referred to in the literature as SMI (Structure of Management

---

[16] Strong arguments have been advanced in some standards groups in the past on the use of set versus action. Some have argued that set allows generic software development versus action. On the other side, arguments have been made that with action you give the semantics of doing a process and this is more conrollable than if it was just a change in an attribute value. Both of these arguments can easily be refuted. When special behavior is associated with modifying an attribute such as checking integrity constraint, special software development will be required. Similarly, action does not give more control (giving only the appearance of control to a modeler) than a set with specific behavior.

[17] This information is not required to be retained for the lifetime of the managed object; it is required only to perform the action request successfully.

Information) operations. (Similarly, the notifications are called SMI notifications.) The operations and notifications specified as part of the information model define what is visible at the boundary of the managed object. In other words, the internal details about the resource are concealed. Mapping between these operations and notifications to a suitable communication protocol such as CMIP is required to transmit what is available at the managed object boundary to an external system.

Table 1.3 illustrates such a mapping when services and protocol of CMISE are used for external communication. The table is specified by mapping to the CMIS services; transforming from the service to the protocol transmitted across an external system/interface may be obtained from the CMIP standard.

TABLE 1.3   MAPPING BETWEEN SMI DEFINITIONS AND CMIS SERVICES

SMI Operation/Notification	CMIS Service
Get	M-GET
Replace	M-SET
Replace with Default	M-SET
Add Member	M-SET
Remove Member	M-SET
Create	M-CREATE
Delete	M-DELETE
Action	M-ACTION*
Notification	M-EVENT REPORT*

*CMIS Services where further syntax specification is required in the context of System Management Functions or resources.

Previous discussions have not addressed the question of how two basic operations prevalent in database—creation and deletion—are incorporated into an information model. The approach taken to include these operations in this paradigm is not in the definition of the managed object class itself; instead, rules are provided on how they are named or unambiguously identified outside of the class definition. This method provides flexibility, and allows creation and deletion at the instance level instead of requiring all instances of a class to follow the same rules. Naming rules, as discussed later, enables new schema to be included after the managed object class definition is completed. With this approach, some instances of the same class may be created by external request and others based on internal resource requirements.

Even though the information model is not used to capture the concept of synchronization, it is worth noting here in the context of operations. Using the management protocol defined for use in TMN, we can specify either best effort or atomic synchronization. The synchronization referred to here is not across multiple objects in different systems; the same managed system is responsible for the multiple objects. Because the requirement to support atomic synchronization is not part of the information model or schema, the support is left to local decision of the managed system (in other words left to implementation). Note that in some cases behavior may be used to specify the need for using atomic synchronization.

1.12. MANAGEMENT INFORMATION BASE (MIB)

The distinction between managed object class and managed object (instance) was discussed earlier. The class definition is part of the information model or the schema. Managed objects that are instantiation of the classes in a system such as an NE form a repository referred to as the "Management Information Base." The actual database used to store the MIB may vary across different implementations and is not relevant for defining the schema. Figure 1.6 shows an example of the MIB[18] concept. Information in the MIB pertains to properties of the various managed resources that can be made visible across an external interface. When the MIB concept is realized in an implementation using various databases, additional information may be included to assist in the internal working of the managed system. These are not visible across the interface.

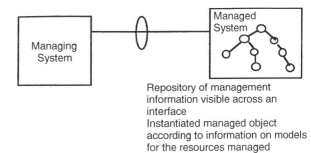

Repository of management
information visible across an
interface
Instantiated managed object
according to information on models
for the resources managed

Figure. **1.6** Management Information Base (MIB) Concept.

The term *MIB* has been used in another context—the approach for managing the data communications equipment (bridges and routers) using the Internet management protocol, Simple Network Management Protocol (SNMP). Modeling principles defined in this approach do not make a distinction between a class representing a type of resource and a specific instance of that class similar to that made earlier in this section. The schema itself is referred to as MIB. Because of the much simpler principles used to define the schema and how the objects (data) are identified, this term is applicable when referencing the schema as well as the repository in an implementation. In other words, an Internet RFC containing the structure of management information is known, for example, as ATM MIB, SONET MIB.

---

[18] The reason why instances are depicted as being in a tree structure will be discussed later in this section. This is the result of how instances are to be named, which is defined as part of the information model.

1.13. EXTENDING MANAGED OBJECT CLASS DEFINITION

One advantage of using the object-oriented design, identified earlier, is reuse at the specification level as well as the implementation level, along with extendibility. When a managed object class is first defined, certain properties are included. Often, one encounters several reasons for extending the original specification to add any of the characteristics mentioned earlier. Extension may be required because new features are now included as a result of enhancements to a service, introduction of new technology, or a supplier interest in providing new features as a market differentiator for their product. The process of extending an existing definition with new properties is referred to as specialization.

Associated with the process of specialization is the concept of inheritance. When a new managed object class is defined by specializing an existing definition, it is said to inherit the properties of the original class. The specialized class is referred to as the subclass, and the original class is called the superclass. Inheritance is a powerful mechanism for building reusable specification. A new managed object class may be defined to inherit from multiple object classes. The restriction on the type of inheritance, regardless of single or multiple, is that it is a strict inheritance. In other words, the subclass includes by inheritance all the properties of the superclass(es) and adds new properties. The collection of managed object classes defined in this manner forms a tree, referred to as class hierarchy.

Figure 1.7 is an example of how inheritance is used to define object classes. The example chosen is taken from the termination point model in Recommendation M.3100.

In Figure 1.7, termination point object class is specialized to form new object classes: trail termination point source and sink. The properties of termination point are applicable regardless of whether the trail termination is a source or sink for signal flow. By defining the generic object class called termination point, the specification is reused, and only additional characteristics for the trail termination source and sink such as downstream and upstream connectivity pointers are included. The specification is flexible and extendible because new subclasses for different technologies can be specified without affecting existing definitions.[19] The figure also includes an example of multiple inheritance when the termination point is bidirectional or when PM data are included for the termination point.

In order to utilize the inheritance concept, it is common practice to define object classes at higher levels of the hierarchy (in contrast to the leaf level) with generic properties. In the above example from TMN, generic source and sink terminations for the transmitted signal are defined, including properties such as states and the

---

[19] This can be also be a disadvantage if sufficient care is not taken because it may result in a proliferation of managed object classes. While it is not a serious concern from the specification point of view, having many flavors of classes with similar functionality makes it difficult for implementation. Unless an implementation can accommodate all the specializations, ability to interoperate will be reduced. This matter is discussed in a later section.

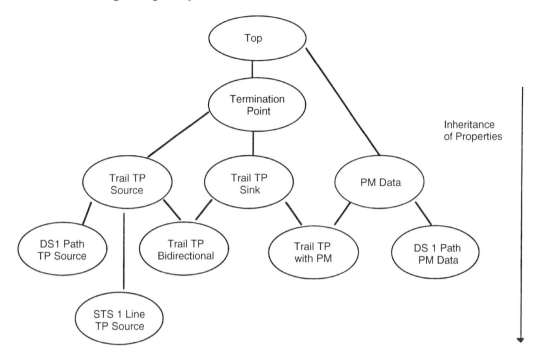

Figure. 1.7 Example of Single and Multiple Inheritance.

connectivity relationship. Using these generic definitions, we define specific subclasses for SDH and ATM. When generic managed object classes are defined, the requirement for strict inheritance should be carefully considered. Including properties that may not be applicable (sometimes it is difficult to know what types of specializations will be required at the time generic definitions are created) for all subclasses will lead to managed objects carrying extra baggage without serving a management purpose.

When defining new object classes using multiple inheritance, care must be taken to ensure that no inconsistency is introduced between characteristics inherited from multiple classes. The subclass definition should resolve these inconsistencies or contradictions. Furthermore, if an attribute, attribute group, notification, or action is included in multiple classes, only one occurrence will be in the managed object of the newly created subclass. However, values restrictions will be combined. As an example, if an attribute in one superclass has the permitted range of values 1..10 and another superclass 1 . . 15, the subclass defined by multiple inheritance from the two classes will permit the range 1 . . 15 for that attribute.

Figure 1.7 is drawn using "top" as the starting object class. All managed object classes defined for the purpose of management are subclasses of top at some level. The definition of top is contained in X.720 and X.721. The properties defined for top are two mandatory attributes that identify the actual class of a managed object and an identifier referencing the naming rule used in instantiating the object. In addition, two conditional packages are included, with each one providing one attribute. One attribute is used to identify the registered[20] packages. The other attribute represents the ability to behave as different object classes.[21] Every managed object class defined as part of TMN includes these characteristics because of inheritance.

1.14. ALLOMORPHISM

Even though allomorphism[22] is not part of the principles used to define a schema, a chapter on information modeling using the principles applied in TMN will not be complete without a brief introduction of this concept. This concept facilitates software release independence to some extent.

Allomorphism enables an instance of a managed object class to be managed as an instance of another class. In order to support this capability, the two classes (the actual class of the managed object and the class used when managing it) must be compatible. Compatibility between two classes are governed by rules described in X.720. Examples of the rules are:

- For conditional packages, the result of evaluating the same condition must be the same regardless of the actual object class of the managed object.
- In order for the managed object to perform the same action included in compatible classes, regardless of its actual class, the mandatory parameters supplied with the request must be the same.

The concept of allomorphism is associated with a managed object. In other words, different implementations of the same class may not exhibit this property. In discussing the properties of top, it was mentioned that a conditional package is defined with an attribute that reflects this property.

---

[20] In order to understand the complete meaning of "registered," knowledge of the mechanism used to provide globally unique identification for the properties is required. For ease of understanding, assume that a package is given a globally unique number under certain conditions. For example, conditional packages are assigned a globally unique value. When a managed object is created, with conditional packages, the values of this attribute identify the packages included in the object.

[21] This is a simple way to describe the complex concept called allomorphism. Further explanation is provided later.

[22] In object-oriented programming, the term *polymorphism* is used to explain how the same function used with different resources will provide what is appropriate for that resource or different methods may be used for the same interface invocation. Even though there are some similarities at some level, allomorphism is different from polymorphism.

Let us now see how this concept facilitates release independence. Even though, in theory[23] a managed object may behave as classes that are not related by inheritance, let us take the case where a circuit pack object with some characteristics is defined in a standard. Suppliers, to meet capabilities unique to their product (not modeled in the standard object class definition), may define a subclass with these additions. Let us suppose that the circuit pack is managed by two systems. Also assume that the software corresponding to the extra capabilities was included in only one managing system when it was incorporated in the managed system. If the circuit pack managed object can exhibit allomorphism, then one managing system can manage using the new class with additional features. However, the other managing system can continue to manage as if the circuit pack were still the standard class. Different releases of the software to manage the same circuit pack can coexist without requiring flash updates in all the systems.

1.15. NAMING MANAGED OBJECTS

Characteristics of the managed resources are described in the schema or information model using the managed object class definition. Given a definition, instances are created to correspond to the resources being managed. When multiple instances of a class are created within the TMN, unambiguous identification of the managed object is required. The scope within which the managed object name is unambiguous may vary from global to relative to the managed system responsible for management of the resource.

In order to meet requirements for both global and local uniqueness, the containment relationship is used. Containment may or may not reflect physical containment. For example, in the case of a circuit pack, it is named relative to the slot in which it is placed. The name of the slot itself will be relative to the shelf and so on. However, a log record may be named relative to the log in which it is entered from a logical perspective, even though it is contained in a disk. Figure 1.8 illustrates naming using containment.

Even though names are pertinent only to instances and not classes, the structure for determining how an instance of a class will be named is specified in terms of classes. The specification of this structure is referred to as "name binding." Name binding expresses how an instance of a class, referred to as the "subordinate class," is named relative to another class, the superior class.[24]

---

[23] The reason for this statement is as follows. When referencing an object with a class other than the actual class, the agent implementation must first recognize the name unambiguously. As we will see later, depending on the naming rule, instances of different classes not related by inheritance often have different naming sequence. In order to successfully implement the behavior rules for allomorphism, the practical approach is when the classes are related by inheritance.

[24] When learning the information modeling concepts discussed here, often confusion arises between the inheritance tree and the naming tree (resulting from the containment relation and name binding rules). Inheritance and naming trees are distinct. Classes defined in name binding as superior and subordinate should not be mixed with superclass/subclass defined as part of the inheritance hierarchy.

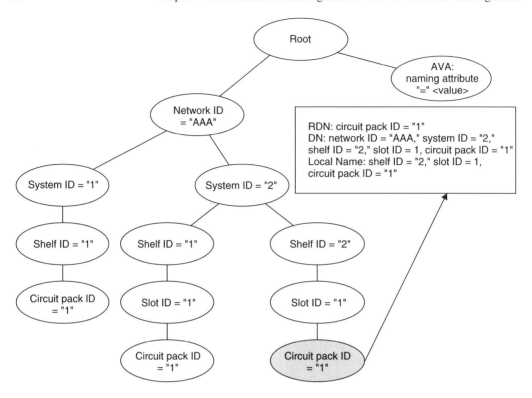

Figure. 1.8 Example Naming Tree.

In addition to identifying the superior and subordinate object classes, the name binding specification also includes the attribute used to name an instance of that class. The value of this naming attribute must be unique across multiple instances of the same class contained in an instance of the superior class. In the figure, slot ID is a naming attribute. A second slot contained in the same shelf (Shelf ID = 2) should have a different value for the slot ID in order to achieve unambiguous identification.

The example in Figure 1.8 also illustrates how to construct local and global names. The local name is relative to the managed element object that represents the NE. Once the managing system establishes communication with the NE, using local names automatically implies that the object is contained relative to that NE. Global name, on the other hand, provides global uniqueness. This is appropriate in cases where a network level management system needs to understand references to objects that may be in different NEs. The global name, however, is long. Another point to recognize here is that the global names are constructed by containing the tree of managed objects relative to objects available from The Directory.[25]

---

[25] The Directory referenced here is the standard X.500 series. The object classes such as organization are defined in X.521.

1.16. MODELING RELATIONSHIPS

The concepts described so far have addressed the properties of the objects. Relationships between objects are expressed in one of three ways: (1) an attribute in one managed object pointing to another managed object, (2) the containment relationship used for naming an instance, and (3) a managed object class that represents the properties of the relationship between two objects. Examples of the three approaches are: equipment that has a pointer to the objects that are affected if there is a failure in that equipment, naming a circuit pack relative to the equipment holder (slot) where it is inserted, and cross connection representing the properties between a from and to end points within a network element that are connected together. Even though relationships could be modeled with the concepts mentioned above, such an approach does not adequately address the specification of a relationship.

Some of the deficiencies are concepts such as role, cardinality, and integrity constraints associated with the relationship itself. These are not explicitly expressed except via behavior statements. To solve this concern, a separate standard was developed (X.725|ISO/IEC 10165-7) to consider relationship modeling known as Generic Relationship Model (GRM). However, the result of such a model should still be expressible across the interface in terms of the managed object concepts mentioned earlier.

In GRM, the approach taken was to consider relationship as an abstraction without being concerned with the actual realization. That is, relationship should be defined regardless of the representation as an attribute, naming relation, or a managed object. The relationship is defined in terms of roles, behavior, relationship management operations, and notifications. The roles themselves are defined in terms of cardinality, various managed object classes participating in the relationship, and how they come into or are removed from the relationship. The managed relationship is defined in terms of relationship classes. Relationships are realized using relationship mapping of the various components of the relationship class in terms of managed object classes, relationship objects, attributes, and operations.

Notational support for defining relationship classes and relationship mapping are provided.[26] In support of reuse in specification, a managed object class called generic relationship object has been defined. This is specialized from "top" mentioned in the previous section. Additional attributes included are: relationship name, relationship class, and relationship mapping.

In TMN, the work in progress for network level models is planning to use this technique. Relationships span managed objects in multiple systems at the network level and above; it is therefore natural to look at this approach as an appropriate technique to use. Without the introduction of GRM, it is difficult to represent information such as role cardinality, permitted operations on the relationship, and how a resource may enter or depart from a relationship.

---

[26] Existing TMN models and ISO standard models do not include the concepts identified by the general relationship model. The first example of its use is in the Domain and Policy management function.

1.17. REPRESENTING INFORMATION MODELS

Using the principles for information modeling defined in previous subsections, we can identify components of the managed resources. However, without a standard representation technique, different designers may choose to specify models in a variety of ways ranging from using a formal description language, pseudocode, or text. This obviously leads to difficulties in interpretation and implementation of the models. As part of the standards, a semiformal[27] technique is defined in ITU Recommendation X.722|ISO/IEC 10165-4. This technique is called Guidelines for Definition of Managed Objects (GDMO). In addition to representing the semantics of the information model using this technique, the syntax of information such as the value of an attribute, parameters sent with a notification for external communication are represented using Abstract Syntax Notation one (ASN.1),[28] defined in the ITU Recommendations X.680 series.[29]

GDMO provides templates to be followed when specifying the various components of an information model: managed object class, package, attribute, attribute group, notification, action, name binding, and parameter.[30] The template consists of key words in describing much of the semantics and hence facilitates parsing by machine. The aspect that is not machine parsable is the behavior description. The behavior is written in text form with the natural consequence of being prone to ambiguity and misinterpretation. Formal Description Techniques such as "Z," "object Z," "SDL," and "LOTOS" may be used instead of text. However, there is no one recommended language to use; some of these languages have been used in the industry for automatic code generation. Contributions were presented in ITU SG 15 working on management of transmission systems using "Z." The main concern with using these languages is the additional knowledge (some of the languages are quite complex) the modelers require to ensure that the description is correct. Support for this approach is therefore not unanimous.

[27] The term *semiformal* is used to indicate that not all parts of the specification technique are machine parsable.

[28] ASN.1 was developed to provide an abstract representation of application-specific information exchanged between two systems. Different encoding rules are applied to the specification using ASN.1 for generating the actual octets transmitted across an interface.

[29] Reference is provided to the revised version of the standard commonly known as X.208. To a large extent X.208 and X.680 (not the others in the series) are identical. X.680, in addition to extensions, incorporates corrections based on implementation experience.

[30] The parameter template is used to specify details when an extension field is included in the syntax. To amplify, the concept of providing extension capability was mentioned in the earlier discussion on notifications, actions, and compatibility. In the syntax of a notification, for example, it is common practice to include a field that provides a hole that may be filled in later if the notification is reused in another managed object class. The parameter template is used to specify the syntax to be used for the hole.

Because this chapter is concerned with information modeling more than with how to represent them, further details are deferred. The reader is directed to the standard or books available in the literature.

1.18. DIFFERENCES IN INFORMATION MODELING PRINCIPLES

Section 3 pointed out that, even though information modeling is used in many applications, the details differ. Some differences are summarized in this subsection.

Information models developed by TMN groups and OMG use object-oriented design concepts such as inheritance and encapsulation. The structure of management defined for data communications network management specifies objects that do not possess these properties. The distinction between an object class and an instance of the class is not present. As a result, multiple inheritance or strict inheritance is not applied when developing the information models. Each object type definition in SNMP can be considered to be roughly the equivalent of attribute definition. A collection of information that needs to be grouped together (similar to an object class) is defined using tables. Differences also exist with regard to how an object is named. Because of lack of distinction between class and instance, the name for the type is also the same for an instance except with table representation. In the latter case, an additional index is used to reference a specific row of the table. Even though ASN.1 syntax is used to represent the information exchanged across an interface, properties of the object (data) are specified using a different notation (ASN.1 macro) instead of GDMO. Some major differences are indicated in Table 1.4.

Information models from OMG define interface types similar to object class. Multiple inheritance (with some differences) is used to form an interface hierarchy/ graph. While the types of interfaces of the objects in TMN and SNMP are for communications, OMG definitions support programmatic interfaces to facilitate application portability.

TABLE 1.4 SUMMARY OF DIFFERENCES BETWEEN INFORMATION MODELING PRINCIPLES

TMN Models	Internet Management Models
Object classes are collections of properties associated with a resource and are reusable.	Object types are atomic data or tables and are not reusable.
Object classes may be specialized using multiple and strict inheritance.	The concept of inheritance is not used.
Object classes may contain optional attributes and coexist with mandatory attributes.	All variables (object types) within an object group (such as a table) are mandatory.
Containment relation is used for naming objects and results in globally unique names.	The concept of containment does not exist, and the name is unique only within a single system.
No restrictions on the ASN.1 types are used for specifying the syntax of the exchanged information.	Only simple ASN.1 constructs and restricted basic types are permitted for defining the syntax.

1.19. EXAMPLES OF INFORMATION MODELS FOR TMN

Having discussed the principles to be used in developing an information model, let us now take a look at some examples of information available from standards for TMN. Before discussing these examples, an overview of the different modeling efforts for TMN is presented.[31]

1.20. TMN MODELING EFFORTS

ITU[32] SG 7 has developed information models to meet the requirements of systems management functions in general. These functions are generic in the sense that they are applicable to management of both components used for data communications as well as telecommunications. Regardless of the technology used to provide a service, resources emit alarms to indicate fault or failure. The generic function "alarm reporting" specifies five types of alarms (equipment, communication, environmental, processing, and quality of service) and information associated with alarms (probable cause, severity, diagnostic information, specific problems, etc.). This definition is used to support the alarm reporting function in the TMN function set "alarm surveillance." The above case is generic from the perspective that these alarms may be associated with resources supporting various technology. Another function modeled is "Event Report Control." In the discussion on notification, it was noted that a resource (managed object) emitting a notification does not imply it will be communicated to an external system. Criteria may be associated with what events should be sent to a particular system. The model to define and apply the criteria is generic regardless of the notification type, TMN function, or resource. This model is discussed in more detail later in this chapter. These generic functions promote one form of reuse. The software for this model, once developed, can be reused for all the TMN functions associated with different resources. For the TMN X[33] interface, the model to support trouble administration function (X.790) was defined incorporating the information models developed in ANSI T1 and NM Forum Omni*Point1* specification.

ITU SG 4 has developed a recommendation on generic network element information model (Rec: M.3100-GNIM). The model at the present time supports superclasses that are technology independent and suitable for both switching[34] and

---

[31] The standards covered here are those from ITU. In North America, standards for X-interface and specialization of ITU Recommendations to meet specific North American needs such as performance monitoring of DS1, DS3 (line and path), and SONET terminations have been developed. Similarly, ETSI has developed specialization for European requirements. Other organizations where information models for use in TMN are defined include ATM Forum, NM Forum, and Bellcore.

[32] Previously known as CCITT.

[33] TMN Standard Rec. M.3010 discusses various interfaces for exchanging management information. Q3 is the interface between a Network Element and OS or between a mediation device and an OS. X interface is between OSs in different administrations.

[34] Support for switching network elements is minimal. The standard was developed using the principles outlined for generic (G.805) and SDH-specific (G.803) transmission architecture.

transmission network elements. Work is in progress to address other levels of abstraction such as network and service. Examples of the managed object classes defined are: managed element (NE), different classes to support point to point, and point to multipoint cross connections, and termination points.

ITU SG 15 has developed information models to support management of SDH and ATM network elements. The SDH models documented in the G.774 series define subclasses specialized from Recommendations M.3100 and Q.822 (Performance Monitoring). A model is also provided at a generic level for protection switching. The model for ATM has been completed recently and is in the final ITU approval process. Part of SG 15 work (network level modeling) has been moved to SG 4 since November 1996.

ITU SG 11 has developed information models to support both specific functions independent of the resource (technology independent) and specific technologies. A service provisioning model provides a framework for administering subscriber information in an ISDN switch. The standard defines superclasses and explains how to model bearer services, supplementary services, optional user facilities for the packet mode, directory number, access port profile (service characteristic associated with an access port), and resource aspects. However, to provision specific services, the framework must be extended.

Models containing the framework for TMN function sets, alarm surveillance, performance monitoring, and traffic management have been completed, using the generic model for performance monitoring and ITU SG 15 developed SDH-specific subclasses. Work is in progress to support the V5 interface (configuration, fault, and performance monitoring functions) and usage metering (call detail record) function. Management models are close to completion for network elements in the SS7 network. Starting in November 1996, TMN interface work in SG 11 has been moved to SG 4.

Three examples are included in this section: (1) a model for event report control function, (2) a cross-connection model, and (3) the framework for performance monitoring. Instead of including GDMO definitions, the models are discussed in terms of how various modeling concepts described earlier are used. Referenced standards should be used if the reader is interested in the formal definitions.

1.21. EVENT REPORT MANAGEMENT

A model for event report management should meet the following requirements:

- Identify either a single or a group (for multi-casting) of destinations where the event[35] should be forwarded via a communication interface.
- Specify criteria such as forward the notification if it is of type communication alarm and the severity is critical or major.

---

[35] The phrases "event" and "notification" are used synonymously. In the standard, notification is used to describe what a managed object emits, and event report is used when the notification is communicated to an external system.

- Schedule the times to forward the notifications.
- Control the initiation, suspension, resumption, and termination of the event report activity.
- Identify the backup destination if communication with the primary destination is not available.
- Configure whether the event report should be communicated requesting confirmation from the managing system.

Except for requirements related to destination and the need for confirming the event report, other requirements are applicable to any activity, including reporting events. In order to provide for reuse, a superclass called discriminator was defined with the properties shown in Figure 1.9.

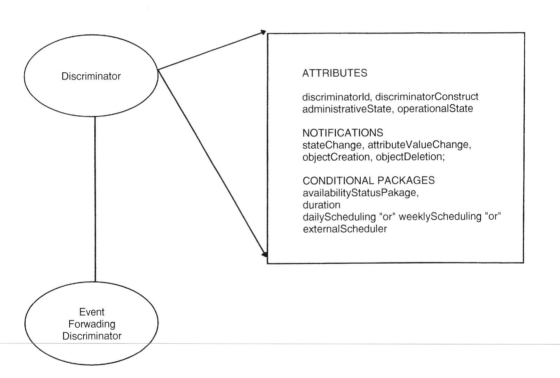

Figure. 1.9 Class Hierarchy for Event Report Control Function.

The inheritance used for discriminator is from "top." In addition to the new attributes defined here, four attributes specified earlier for "top" are also present. The semantics of the components defined for the discriminator managed object class are as follows:

discriminatorID: This naming attribute is used to identify uniquely an instance of either the discriminator class or its subclasses relative to the containing object.[36] The value is settable at creation time.

discriminatorConstruct: This attribute specifies the criteria to perform an activity. It is an expression defined using logical operators. In order to perform the activity, the criteria must evaluate to true. An example of defining a criteria is "perform this activity if (attribute x > 5 and attribute y is not present or attribute y has an initial string equal to "sri") is true." The logical expression may be modified by the managing system.

administrativeState: By setting this state attribute to locked or unlocked, the service offered by the discriminator is suspended or resumed. (Initiation is equivalent of resumption if at creation time the value is unlocked.) The value is settable by an external system

operationalState:[37] This attribute reflects the operability of the managed object. The values are enabled or disabled. The values are not controllable by an external system. The value of the state is changed internally based on whether or not the service offered by the discriminator is available.

stateChange, attributeValueChange, objectCreation, and objectDeletion notifications: These correspond to events that are generated as a result of changes in the values of the attributes, changes in states and status, and creation and deletion of the discriminator, respectively. Note that even though states are also attributes, separate notifications are defined to distinguish from changes to any attribute in general. State, status attributes generally applicable across multiple objects, and state change notification are defined in ITU Recommendation X.731, X.721|ISO/IEC 10164-2, 10165-2.

Conditional packages for scheduling: Availability status, included in a package, is used to indicate whether the discriminator activity is scheduled on or off. Different types of scheduling are specified: the duration is intended to provide the start time (which may be the time when the object is created) when the services of the discriminator are made available and an end time (includes continual availability as long as the object is in existence). Within the duration, one can schedule on a daily basis, for example, availability only between 8–5 every day or for each day of the week, or on a per week basis (for Mon–Fri 8–5 and not available on Saturday and Sunday). Instead, an external scheduler such as the ones defined in ISO/IEC 10164-15 |X.746 may also be used. The external scheduler has the advantage that the same scheduler may trigger activities in multiple objects (same or different classes), whereas the daily and weekly scheduling packages affect only the object containing them.

---

[36] TMN specifications define a managed object class called managed element to represent a network element. This is used as the superior object for discriminator and its subclasses.

[37] For an object like a discriminator, the semantics of operational state may be difficult to comprehend. This is an object that represents the logic used to program this function. On the other hand, it is more natural to understand that the operational state of a circuit pack will change to disabled when the card fails.

The event forwarding discriminator (EFD) is specialized from discriminator, with the special behavior that the activity performed relates to forwarding events to one or more destination. The EFD object class, in addition to this enhanced behavior, adds the mandatory attribute destination. Destination is defined to be either one application entity (an application in a specific system) or a group of entities. The latter case supports the broadcasting environment.

The following two conditional packages are included:

- backUpDestination: This package includes two attributes—a list of destinations to be used in the given order if communication to the system identified by the destination attribute fails,[38] and the attribute active destination to specify the currently active one based on what has been selected from the backup list. Because the active destination is set as a result of selecting one of the items in the backup list, only a read operation is permitted.
- mode: This package, if present, facilitates a managing system to configure the method of receiving event reports. The protocol CMIP specifies two methods of issuing event reports. In the confirmed mode, the manager sends a response acknowledging receipt of the event report. By setting the value of mode attribute to confirmed or otherwise, the managing system can define how the event reports should be issued. If this is not present, then the decision is local to the managed system.

The model for event report management forwards all events to all EFDs.[39] Depending on the result of evaluating the criteria, an event may or may not be forwarded to a specific destination. The standard defines the concept of a preprocessing function. The syntax for a notification may include information that is to be determined by a process outside of the managed object itself. For example, if an alarm is issued with a minor severity and later with a major severity, indication that the two alarms are correlated is done by the preprocessing function prior to including it in the event report. The standard also accounts for an EFD that is not manageable, in other words, the logic for determining which events should be forwarded is internal to the managed system and is not configurable by the managing system.

---

[38] New work is in progress for an enhanced event report control function. This introduces a new object class disseminator in which the notifications that could not be issued because of communication failure are queued and disseminated later when the links are set up again.

[39] Even though, conceptually, all notifications are seen by all EFDs, in practice this is not required to be implemented in this manner. Depending on the number of EFDs and the criteria to be checked, this may not be optimum for software development. In some cases, specific name bindings are provided to restrict the above behavior so that program logic can be simplified.

1.22. CROSS-CONNECTION MODEL

The model to support cross-connection assignments was developed as part of the Generic Network Information model in ITU Recommendation M.3100. The requirements to be met include:

- Assign a particular channel or time slot for specific use (designation as active Embedded Operations Channel to support access arrangements between a switch and an access node).
- Consider cross connecting a group of terminations with a certain bandwidth used to carry services with another group of terminations with the same bandwidth.
- Provide one or more point-to-point or point-to-multipoint cross connection with a single request.
- Cross connect a termination point selected from an available pool to another termination (either a specific one or selected from a pool).
- Create pre-provisioned cross connection in a state where traffic will not flow until later (by turning on with a management request).
- Create the cross connection as uni- or bidirectional.
- Trace connectivity within the network element when flexible cross connections are present.

In addition to the above generic requirements, within North America when a cross connection is part of a sensitive circuit (line to the President), it is identified as a "red-lined cross connection". The ability to create these specialized cross connections should be supported in some administrations. Details on how to use the generic cross-connection model and examples of its application in SDH (using termination points defined in G.774) are described in an annex to M.3100 (1995).

The object class hierarchy is shown in Figure 1.10. Managed object class fabric is responsible for receiving requests to create and delete cross connections. In addition, it supports actions to create and modify the pool of available terminations and group of terminations for concatenated payloads. The original definition of fabric developed in 1992 was further specialized in the revised Rec. M.3100 to support switching one of the end points of the cross connections to a different end point (termination).

In addition to the naming and state attributes of fabric, the fabric model defines the following actions:

- Add termination point(s) to a group termination point (used for concatenated pay loads).
- Remove termination point(s) from a group termination point.
- Add termination point(s) to a pool.
- Remove termination point(s) from a pool.
- Connect to create one or more cross connections.
- Disconnect to delete cross connection(s).

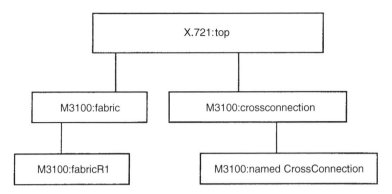

Figure. 1.10 Cross-Connection Model Class Hierarchy.

Cross-connection object definition defines state attributes, from and to termination points of the cross connection, and directionality to indicate uni- or bidirectional flow and signal type. The definition of multipoint cross-connections specifies only the from termination. The "to terminations" are determined from the cross-connection objects contained in the multipoint cross-connection object.

The cross connections created are named as shown in Figure 1.11 relative to the fabric that was requested to create the cross connection. In the case of point to multipoint, the multipoint cross-connection object is contained in the fabric and, as stated above, contains the various cross-connections to the multiple to terminations. Figure 1.11 also shows that when a group termination point (GTP) or TP pool (pool of terminations) is created, it is contained relative to the fabric.

One of the requirements is to include enough information to facilitate tracing connectivity. This is seen in Figures 1.11 and 1.12. The cross-connection objects

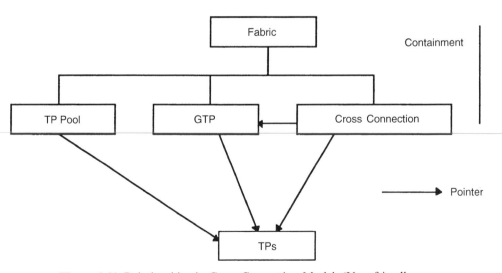

Figure. 1.11 Relationships in Cross-Connection Model. (User-friendly
names, instead of M3100 names, are used.)

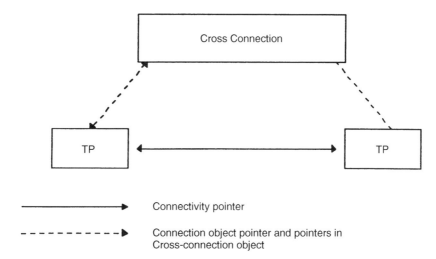

Figure. 1.12 Connectivity Relationships Between Termination Points.

(simple point to point case) have pointers used to identify the connectivity between the terminations via the cross-connection object. The termination point model also supports the backward relation by pointing to the cross-connection object. When there is no flexible cross connection, the two termination points forming the connection will point to each other.

Figure 1.13 illustrates the use of cross connection in an integrated digital loop carrier system. This figure is taken from the model defined in Bellcore GR 836 (defines termination points shown in the figure subclassed from ITU Rec. M.3100) and GR 2833 (to support fiber in the loop, IDLC architectures). The cross connection shown is unidirectional and connects a ds0 within a ds1 to a ds0 (contained within the analog line termination) on the customer side. The use of pointers is quite excessive; however, the complete connectivity information can be traced.

1.23. PERFORMANCE MONITORING FRAMEWORK

The framework for performance monitoring model was developed by ITU SG 11 and is documented in Recommendation Q.822. This is a framework because the managed object classes defined in Q.822 for collecting performance monitoring information are not implementable without additional technology-specific information. In addition to the requirements resulting from the definition of TMN performance monitoring functions (scheduling data collection, reporting on threshold crossing, etc.), examples of the criteria considered in developing the model include:

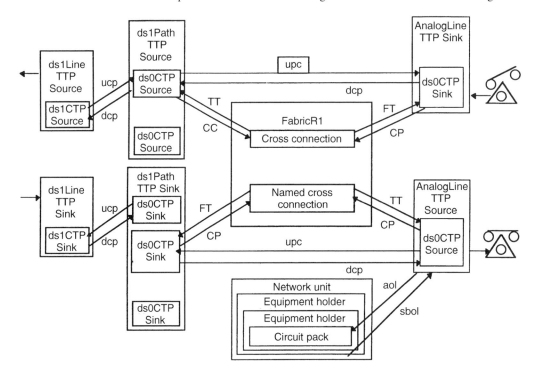

Figure. 1.13 Application of Cross-Connection Model in IDLC System.

- Flexibility to add new performance parameters in two cases: (a) new capability available in the resource, and (b) add or remove dynamically parameters for collecting data for a specific interval.
- Addition or removal of the collected PM parameter values without affecting the existence of the managed object representing the monitored resource.
- Ability to retrieve history information pertaining to individual parameter values.
- Storing object threshold values of the PM parameters applicable to multiple managed resources.

Figure 1.14 shows the managed object class hierarchy for the PM framework. This framework has also been used in the traffic management model. The figure includes both the framework and an example of the specialization required for specific technologies. The scanner object is defined as part of the Metric Objects and attributes (used in generating statistical information), with the behavior to scan and report the scanned data. In addition, scanner includes attributes such as granularity period, which determines how often the parameters are scanned, and attributes for scheduling the scanner activity. The subclass current data include the mandatory

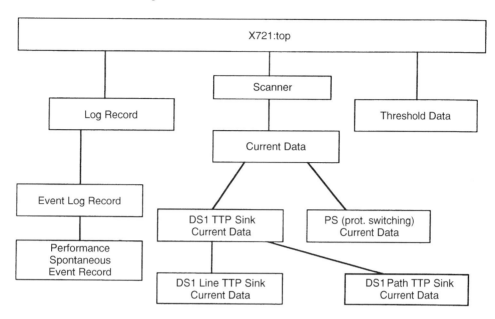

Figure. 1.14 Performance Monitoring Class Hierarchy.

attributes to represent the elapsed time during the collection period and a flag to indicate the validity of the collected data.

Several conditional packages are included. To support the requirement for adding or removing new parameters, the measurement list attribute in the conditional package is used. Because this is a list of attributes, even after the object is created, additional parameters that require collection for special-purpose analysis can be included. Another package of interest contains the attribute that specifies how long the collected information must be retained. In performance monitoring, if the resource is working properly, many of the performance parameters will have a value of zero. Instead of recording several intervals with zero values, using the zero suppression package, it is possible to suppress them, thus avoiding records with just zero values. Two conditional packages are defined to contain two notifications. One notification defines the structure for reporting PM data collected at the end of each interval. The second notification is a quality of service alarm corresponding to threshold crossing alerts.

The instantiable managed object classes are defined by subclassing current data. The subclass includes attributes corresponding to the performance parameters appropriate to that technology. In Figure 1.14, the DS1TTP sink current data is the superclass that contains performance parameters common to both path and line.[40] Further subclasses add parameters that are specific only to line or path.

---

[40] Recent work in ANSI T1 for SONET and PDH has shown that splitting of line and path current data objects in terms of two classes each (for near and far end performance monitoring parameters) is required.

Several subclasses for collecting PM parameters on SDH-specific terminations are defined in G.774.03.

Based on the attribute in the history retention package (defined in the superclass current data), the collected data will be retained in an instance of the appropriate subclass of history data. The threshold data-managed object class is used to include threshold values for the PM parameters. The collected values are compared against the value in the appropriate threshold data object (if present) to detect a threshold-crossing condition.

The notifications may be logged, and two specific log record object classes are defined corresponding to the two notifications.

The flexibility requirement is met by using the containment relation between the current data (actually instantiated subclasses) and the monitored object. For example, a ds1line current data object will be contained in a ds1line object. By using containment, PM parameters are separated from the observed object itself. This implies that new current data objects (PM parameter collections) can be created and deleted without affecting the observed object. In addition, creating subclasses of an already defined technology specific current data makes it modular without having to redefine the observed object. (This would have been required if the PM parameters were defined to be part of the managed object class representing the monitored resource.) However, the number of managed objects is increased to support this flexibility.

1.24. INFORMATION MODELS IN STANDARDS

As mentioned earlier, several information models have been developed or are in progress to support management of telecommunications network elements. A list of models available in ITU and ISO Recommendations are provided in Table 1.5. The table provides the various ITU Recommendation, status, and a brief description of the areas addressed. For work in progress, the status, may have changed by the time of publication of this book. Readers are encouraged to check information available in the public domain such as the World Wide Web.

TABLE 1.5   LIST OF INFORMATION MODELS IN STANDARDS

Document Number (ITU/ISO/IEC)	Title	Status	Description
X.721/10165-2	Structure of Information— Part 2: Definition of Management Information	IS	Definitions to support OSI Systems Management function such as event report control, log control. In addition, generic object classes top, system are included.

TABLE 1.5 *Continued*

Document Number (ITU/ISO/IEC)	Title	Status	Description
X.734/10164-5	Systems Management— Part 5: Event Report Management Function	IS	Model in text for event report control function.
X.735/10164-6	Systems Management— Part 6: Log Control Function	IS	Model (in text) for log control function.
X.746/10164-15	Systems Management— Part 15: Scheduling Function	IS	Definitions of different types of schedulers to allow periodic, daily, and monthly scheduling of an activity.
X.730/10164-1	Systems Management— Part 1: Object Management Function	IS	Model (in text) for managing creation, deletion, changing values of attributes along with generic notifications applicable to any managed resource.
X.731/10164-2	Systems Management— Part 2: State Management Function	IS	Textual description of state model, attributes and notification applicable in general to several managed resources.
X.732/10164-3	Systems Management— Part 3: Attributes for Representing Relationships	IS	Textual description of relationship attributes and notification applicable in general to several managed objects.
X.750/10164-16	Systems Management— Part 16: Management Knowledge Management Function	DIS	Model and definitions to discover the schema implemented in the managed system.
X.751/10164-17	Systems Management— Part 17: Change Over Function	IS	Model and definitions to support the changeover between the active/standby or backup/backed-up relation between managed objects (resources).
X.744/10164-18	Systems Management— Part 18: Software Management Function	IS	Model and definitions to support software activation, deactivation, and interactive aspects of software download.
X.749/10164-19.2	Systems Management— Part 19: Management Domains and Management Policy Function	DIS	Model and definitions to support identifying domains and policies to be applied for management.

TABLE 1.5   *Continued*

Document Number (ITU/ISO/IEC)	Title	Status	Description
X.743/10164-20	Systems Management— Part 20: Time Management Function	DIS	Model and definitions for managing time of day synchronization and accuracy.
X.733/10164-4	Systems Management— Part 4: Alarm Reporting Function	IS	Model in text of the five types of alarms and the information associated with them. Applicable to several managed resources.
X.745/10164-12	Systems Management— Part 12: Test Management Function	IS	A general framework and generic definitions for testing.
X.737/10164-14.2	Systems Management— Part 14: Confidence and Diagnostic Test Categories	IS	Definitions of specific test categories using the framework mentioned in the previous row.
X.739/10164-11	Systems Management— Part 11: Metric Objects and Attributes	IS	Model and definitions of various metering monitors to sample an attribute value over time and calculate statistics such as mean, variance, and percentile.
X.738/10164-13	Systems Management— Part 13: Summarization Function	IS	Model and definitions to scan attribute values from several objects for specific time periods and provide one packaged report.
X.742/10164-10	Systems Management— Part 10: Usage Metering Function	IS	Model and definitions for a framework to collect usage measurements from resources and report according to triggers.
X.736/10164-7	Systems Management— Part 7: Security Alarm Reporting Function	IS	Model (in text) of the different types of security alarms and the associated parameters. Applicable to several managed resources.
X.737/10164-8	Systems Management— Part 8: Security Audit Trail Function	IS	Model and definitions of the information logged to facilitate auditing the security violations.

TABLE 1.5   *Continued*

Document Number (ITU/ISO/IEC)	Title	Status	Description
X.741/10164-9	Systems Management— Part 9: Objects and Attributes for Access Control	IS	Model and definitions to manage the security information used for controlling access to managed resources.
M.3100	Generic Network Information Model	Approved Rec.	Generic Network Information Model to support transmission and switching network elements (concentration is on transmission NEs).
G.774	Synchronous Digital Hierarchy (SDH) Management Information Model	Approved Rec.	SDH specific network element model based on M.3100 and G.803, transmission architecture.
G.774.01	Synchronous Digital Hierarchy (SDH) Management Information Model for Performance Monitoring	Approved. Rec.	SDH specific definitions based on Q.822 to support performance monitoring of SDH NE.
G.774.03	Synchronous Digital Hierarchy (SDH) Management Information Model for MS Protection Switching	Approved Rec.	Generic and SDH specific information model to support different types of protection switching arrangements.
G.774.04	Synchronous Digital Hierarchy (SDH) Management Information Model for Connection Protection	Approved Rec.	
G.774.02	Synchronous Digital Hierarchy (SDH) Management Information Model for Configuration of Payload Structure	Approved Rec.	Information model for the payload configuration management of SDH networks. The functions addressed are used to configure various SDH adaptation functions.
Q.751.1-MTP	Network Element Manager Information Model for Message Transfer Part (MTP)	Approved Rec.	

TABLE 1.5 *Continued*

Document Number (ITU/ISO/IEC)	Title	Status	Description
Q.751.2-SCCP	Network Element Manager Information Model for the Signaling Connection Control Part (SCCP)	In Res. 1 Procedure	
Q.821	Stage 2 and Stage 3 Description for the Q3 Interface—Alarm Surveillance	Approved Rec.	Generic objects to support alarm surveillance function set. Uses X.733 alarm reporting definitions.
Q.822	Stage 1, Stage 2, and Stage 3 Description for the Q3 Interface—Performance Management	Approved Rec.	Framework for collecting and reporting performance management data—applicable to both performance monitoring and traffic management.
Q.823	Functional Specification for Traffic Management	Submitted for final approval	Model to surveil, audit, traffic data from circuit switches and SS7 network elements and for different types of controls.
Q.824.0	Stage 2 and Stage 3 Description for the Q3 Interface—Customer Administration—Common Information	Approved Rec.	Framework model to support provisioning analog and ISDN services. Expected to form the basis for other technologies.
Q.824.1	Stage 2 and Stage 3 Description for the Q3 Interface—Customer Administration—Integrated Services Digital Network (ISDN)—Basic and Primary Rate Access	Approved Rec.	Specialized from the general framework above, model for service aspects as well as resource aspects to provision both basic rate and primary rate interfaces.
Q.824.2	Stage 2 and Stage 3 Description for the Q3 Interface—Customer Administration—Integrated Services Digital Network (ISDN)—Supplementary Services	Approved Rec.	Framework for developing supplementary services. Examples of how to use the framework to support specific services included.
Q.824.3	Stage 2 and Stage 3 Description for the Q3 Interface—Customer Administration—Integrated Services Digital Network (ISDN)—ISDN Optional User Facilities.	Approved Rec.	Model to support packet mode bearer service.

TABLE 1.5   *Continued*

Document Number (ITU/ISO/IEC)	Title	Status	Description
Q.824.4	Stage 2 and Stage 3 Description for the Q3 Interface—Customer Administration—Integrated Services Digital Network (ISDN)—ISDN Tele Services	Approved Rec.	Model to support teleservices in ITU Recs.
Q.942	Stage 2 and Stage 3 Description for the Q3 Interface—Customer Administration—Integrated Services Digital Network (ISDN)—Service Profile Verification and Service Profile Management	In progress	Model based on the generic framework of Q.824.0 and service descriptions for the switch to CPE interface.

IS—International Standard Status
DIS—Draft International Standard Status

1.25. EXAMPLE INFORMATION MODELS FOR DATA COMMUNICATIONS

The information models developed by Internet groups for managing data communications resources fall into two primary categories:

- **Request for Comments (RFC) models:** These are Internet "standards" that have been formally approved by the Internet Advisory Board (IAB). They are developed under the Internet Engineering Task Force (IETF) by less formal working groups than the national (ANSI) or international (ISO and ITU) standards groups, and consist primarily of individual technical contributors rather than national body representatives.
- **Enterprise-specific models:** These are developed by individual contributors and, therefore, are not formally approved. They may be made available to the Internet community as public domain or may remain proprietary.

The initial set of objects to manage TCP/IP networks were developed in MIB-I [RFC 1155]. Since then, the use of MIB-I has been deprecated with the introduction of MIB-II [RFC 1213]. MIB-II is an updated version of MIB-I with the introduction of several new object groups, modification of some variables, and deprecation of one object group for address translation. MIB-II object group definitions include system, interfaces, Internet Protocol (IP), transmission, transmission control protocol (TCP), and SNMP. Enterprise-specific information can be added to MIB-II by adding new variables. Other MIBs available in the public domain include DS1 [RFC 1406], DS3 [RFC 1407], SMDS Interface Protocol [RFC 1304], and others. Internet RFCs are also available for customer network management of SONET and ATM.

The system group is a collection of object types, each being an atomic data. These are distinguished from the example below where a group of properties are collected together in a table. Information modeled as part of the system group are: system description, system object identifier, system up time, system contact, system name, system location, and system service. The access definition indicates whether a specific information may be only monitored or modified. System up time, for example, cannot be modified, whereas changes can be made by management exchange to system contact. These access properties are similar to those discussed for TMN models in section 5, even though there are fewer allowed attribute operations than with the TMN paradigm.

For the IP group, object types include tables to store IP addresses, IP routing tables, and IP address translation tables. An IP routing table is composed of Route entries. Each entry has the following columns: IP Route Address, IP Route Index, IP Route Metrics 1 to 4, IP Route Next Hop, IP Route Type, IP Route Protocol, IP Route Age, and IP Route Mask. The information model for this group further defines representation of these entries. Management of this information is obtained by reading or writing values for a row of this table. Each column is defined as an object type, and a table is composed of these object types. A table, similar to a managed object class in TMN, includes a set of properties; however, it cannot be specialized to add another column. Referencing a row is using an index relative to the table reference.

All the object types within a MIB are mandatory for an implementation. The conditional packages that allowed for optionality in TMN information models are not available. Having no optionality, as can be seen from the next section, results in simpler and interoperable interfaces.

1.26. CONFORMANCE AND INTEROPERABILITY

The major objective of the information models for network management is to promote interoperability in a multisupplier communications network and thus enable efficient management of that network. The previous sections discussed how information models to enable unambiguous interpretation of the management information exchanged between the managing and managed system are defined. From the examples presented, it quickly becomes obvious that reasons such as compromises between members developing the standard, differences in how various administrations set up their network and offer services, and flexibility of the model introduce options in the model. For example, conditional packages with the present condition "if an instance supports it" is nothing more than leaving it an option for the equipment provider to implement the feature. Regardless of whether the communication protocol or management information exchange (application level) is considered, options translate to potential issues with interoperability. This has been demonstrated during the early implementations of X.25 where different options were chosen by different implementors.

It is important to understand the differences between conformance and interoperability. ISO has developed a framework and methodology for conformance testing as part of the ISO 9646 series. An implementor can claim conformance by identifying the options chosen and any restrictions to value ranges, syntax, and so on. These statements from the implementors are tested by organizations such as the Corporation for Open Systems (COS) with a well-defined set of test suites and certify whether the product passes the conformance requirements and check for validity of the claims in the statements from the supplier. Two suppliers may have products that are certified. However, except for the features mandated by the protocol or information model, variations may be found among the various suppliers on the actual options selected.

Let us illustrate the difference between the two concepts using the event forwarding discriminator discussed in section 5. Let us assume that the managed system is implemented without the mode package that allows the manager to configure the EFD regarding receiving notifications as confirmed or nonconfirmed. If an OS sends a request to create an EFD, including the mode attribute, it will fail because the managed system did not implement it. Even though the managed system can respond with the error of unsupported attribute and the managing system can resend the create without the mode attribute, this exchange implies that there is an interoperability issue. Another example is restrictions on the value ranges. Supporting different value ranges by the two communicating partners also leads to interoperability issues.

1.27. CONFORMANCE STATEMENTS

The protocol requirements for the TMN interfaces are specified by selecting options available in the international standards. As part of the international standards specifying the protocols, Protocol Implementation Conformance Statements (PICS) are defined. The conformance statements reflect in a tabular form for every protocol data unit (PDU) which parameters are to be supported and which are optional. In addition, a comment or description column is used to provide any value restrictions and other additional information. These statements, defined as part of the protocol standards, are referred to as static conformance. In other words, the implementor claims by filling the support column what parameters have been implemented relative to the PICS requirements. Implementing a parameter does not imply that it will be sent or received in every exchange of the message.

This is referred to as dynamic conformance, which will be discussed in the section on Profiles.

In addition to the framework and testing methodology, a notation known as Tree and Tabular Combined Notation (TTCN) has been developed to assist in the specification of abstract test suites.

Similar to PICS for protocol, conformance statements are also provided for the information models. The framework for writing conformance statements to the various components of an information model (managed object class, attribute, notification, action, name binding, parameter) is defined in ITU Rec. X.724 |ISO/IEC 10165-6, "Requirements and Guidelines for Implementation Conformance Statement Proforma Associated with OSI Management." The various statements (Management Conformance Summary—MCS, Managed Object Conformance Statement—MOCS, Management Information Definition Statement—MIDS, Managed Relationship Conformance Statement—MRCS) may be used by the implementors to indicate to their customer what is available in their product. The set of conformance statements defined by the above standard addresses conformance from the perspective of the managed system. An amendment is close to final approval for stating conformance from the managing system view.

These statements are specified in a tabular form with the goal of making them machine readable. It is therefore possible to automate the comparison of the conformance statements from the suppliers of the managing and managed system and to determine potential problems with interoperability.

1.28. PROFILES AND INTEROPERABILITY

To facilitate interoperability between suppliers of OSI Standards, the concept of International Standardized Profiles was introduced by the OSI implementors' workshops in North America, Europe, and Asia. ISO developed technical report for the framework and taxonomy of International Standardized Profiles (ISP) in TR 10000-1.

The major goal of the ISPs is to increase the probability of interworking between products from different suppliers. In specifying the profiles, consideration should be given not only to the static aspect of conformance mentioned earlier but also to the dynamic aspects related to the communications exchange. As an example, a parameter within a Protocol Data Unit (PDU) may be defined to be optional. If the static conformance specifies that the parameter is mandatory, this implies that the product must implement that parameter. However, if the protocol defines this parameter to be optionally present in a PDU, then the dynamic conformance will be made optional. In addition, differences may exist relative to sending versus receiving the PDU. A parameter defined as optional may not be present in every exchange of that PDU. However, the receiver must be capable of receiving it if it is sent in the PDU.

1.28.1. Network Management Profiles

The concept of an A-profile has been introduced in the industry by standards and implementation groups. This refers to the requirements on the protocols for the OSI layers 5 through 7. For network management, the requirements for session, presentation, and ACSE are documented in ISP 11183 Part 1. Two profiles called AOM11 (ISP 11183 Parts 1 and 3) and AOM12 (ISP 11183 Parts 1 and 2), are developed for CMISE to support network management. These profiles developed by the implementors' workshops in the United States, Europe and Asia are approved international standards.

It was pointed out earlier that the protocol requirements for the Q3 and X interface are being revised. Requirements in the revised recommendations are specified using the network management profiles.

In addition to the protocol profiles, profiles for each system management function (e.g., log control, event report management) are standardized in ISO 12059 parts. Using these profiles as building blocks, we can define a set of profiles in ISP 12060 parts. For example, ISP 12060-2 includes both alarm reporting and state management capabilities.

The above-mentioned profiles are generic functions that are commonly defined between ISO and ITU and used in TMN. Similar profiles do not exist for TMN applications. Efforts are underway in various organizations to begin work on TMN profiles. The delay is because the conformance statements are not available for all the models in TMN. At the time of writing this chapter, conformance statements exist only for the Generic Network Information Model (Rec. .3100) in ITU Rec. M.3101.

1.28.2. Information Model Ensembles

The concept of ensembles was first introduced in the NM Forum to specify the collection of components of the information models required to meet a specific function. The standards for Q3 interface include management information that is suitable to different types of NEs (e.g., ADM, DCX) and different functions (e.g., cross connection). Ensembles define the collection of objects that are necessary to support different network element functions. (Note that the collection should be specified in terms of logical functions performed by the NEs to allow variations in vendor products that combine different functionalities in a product.) Recently, the concept of solution sets has been introduced by the NM Forum. The solution sets may be generic, such as alarm monitoring, or specific, such as LAN alarm interface. The solution sets provide a high-level description of the problem to be solved, along with references to appropriate documents from regional/international standards and NM forum for the details. All these efforts are aimed at facilitating service providers to request interoperable products from various suppliers without requiring detailed knowledge of how the management information is modeled or the functions are supported by each standard.

1.29. CONSIDERATIONS FOR INTEROPERABLE TMN INTERFACES

The following can be considered a checklist or steps (not necessarily in sequence) required for implementing an interoperable CMISE based interface.

Agreement on the minimum functionality to be deployed in the network to support operations for a specific domain such as alarm surveillance and performance monitoring.

Agreements on the protocol features in CMISE as well as in the lower layers required in order to achieve the above functionality.

Subdivision of the minimum functionality required to support operations in terms of atomic units.

Agreement on the application context to be used.

Subset of the schema required when managing a specific type of NE supporting a specific technology.

Determination of the administration-specific requirements for the selected schema.

Selection of the structure rule for naming instances of the selected object classes.

Security requirements may also have to be addressed in interfaces such as customer network management and between different administrations.

1.30. FUTURE DIRECTIONS

The information models developed in support of TMN addresses the interface between two systems. This emphasis is appropriate for the initial goal of improving interoperability when the network includes multisupplier products. In addition to improving interoperability using standard interfaces, recent advances in software engineering on distributed processing are starting to influence future directions. Some technical reasons[41] for promoting the move toward the new concepts are discussed below. Before introducing these new directions, let us first look at some of the concerns expressed by implementors in building TMN applications based on current specifications.

An important time-to-value consideration in building and deploying products adhering to TMN standards is the availability of the infrastructure components required to easily build TMN applications. The standards do not address implemen-

---

[41] *Soap box*: As will be seen later, new notational techniques are being introduced in some cases, in addition to describing requirements and information models using new terminology from distributed processing. Based on the work in open distributed management architecture, the need for these complex notations (even though the final product is expected to be GDMO-based information models) is very questionable.

tation aspects such as interobject interactions, moving management information across multiple nodes, replication of data for redundancy, and portability of applications using programming interfaces and recovery scenarios. The existing TMN standards considered these issues as being specifically outside the scope of the work. Until recently sufficient tools have not been available in the marketplace to aid the implementors in developing TMN applications without becoming versatile with many new concepts (e.g., different protocols, syntax transformation, object-oriented concepts, relation between managed objects and resources).

To assist in the rapid development of network management products using CMISE and associated information models, specifications providing bindings to programming languages C and C++ have been (are being) developed by the Open Software Foundation and X/Open committees. Platform products are now available (though limited) with infrastructure components such as

- Application programming interfaces
- Processes for database manipulation
- Generation of transfer syntax
- Generation of appropriate communication protocol data units

These components follow the specifications from X/Open, thus facilitating the portability of applications developed using these components.

1.31. DISTRIBUTED PROCESSING AND TMN

While the development of the tools aid in implementations, it should be recognized that the current specifications are based on the fact that the managed resources and management activity are within a single system. The advent of distributed processing concepts such as client/server has produced several benefits. Some of the advantages are load balancing, failure resilience, increased reliability by redundancy, and increased performance because of concurrent executions. Building network management as a distributed application allows distribution of managed resources across multiple network nodes and makes use of the aforementioned benefits. However, this is not a panacea. As always, there is a price to pay to reap these benefits. Examples of the challenges to be resolved are:

- Connection of disparate systems
- Management of partial failure or differences in availability schedules
- Access and location transparency for the information
- Federation of administrative and technology domains
- Security concerns
- Need for clock synchronization

In addition to the above-mentioned general issues for any distributed processing application, additional problems to be solved specific to TMN include:

- Determination of the agent responsible for specific resources
- Global naming of managed objects to handle location transparency
- Maintaining integrity constraints between objects distributed in different systems
- Correlation of distributed management activities
- Migration of existing information models into a distributed environment with minimal adaptation

A first step toward solving some of the issues for network management is the new work on Open Distributed Management Architecture (ODMA). Before describing ODMA, let us look at some of the concepts developed for open distributed processing (ODP).

1.32. OPEN DISTRIBUTED PROCESSING

The open distributed processing standards (X.901-903|ISO/IEC 10746-1 to 4) are being progressed jointly between ISO and ITU. The ODP reference model includes concepts, modeling approach, and levels of abstractions required in building a distributed system. In addition, an object-oriented framework[42] has been adopted for modeling some of the specifications. The five levels of abstractions discussed are:

Enterprise viewpoint: requirements that address the business goals, policies, and environment for the system independent of how the system is distributed.

Information viewpoint: semantic information that should be stored and processed in the system along with the information flow between the source and sink for the information.

Computation viewpoint: further details on how the information can be decomposed into objects that interact via interfaces. The components are defined to meet the requirements for implementing a distributed system.

Engineering viewpoint: realization of the computational model in a specific environment, for example, using specific protocol mechanisms. Furthermore, mechanisms to support various transparencies associated with distribution and infrastructure components are provided.

---

[42] The principles described earlier to define a management information model, though object-oriented, are not exactly the same in ODP.

Technology viewpoint: implementation details of the components required for building distributed systems. This viewpoint is considered to be outside the scope of standardization.

Various distribution transparencies are discussed within the ODP standards. Not all transparencies will be applicable for all applications. Examples of transparencies discussed are:

- Access transparency to mask the variations in data representation and invocation mechanism (for example, regardless of the protocol used to invoke an operation).
- Failure transparency to hide the failure of the object from itself (provides for building resilient systems).
- Location transparency to shield from an object the exact locations of the objects with which it interacts.
- Persistence transparency to mask the activation and deactivation of an object so that it appears to be always present for interaction.
- Transaction transparency to hide the coordination (scheduling, monitoring, and recovery functions) of activities across multiple objects to achieve data consistency.

In addition to the prescriptive and descriptive models, infrastructure components to support some of the above-mentioned transparencies are in progress. These components allow users to obtain information about the available services and access them. One such component defined is the "trading" function. This allows servers to advertise the services offered and clients to discover them, thus decoupling the clients and servers. Other components include the distributed object manager to bind and initiate invocations of services provided by server objects, binder, access function, and type manager.

1.33. OPEN DISTRIBUTED MANAGEMENT ARCHITECTURE

ITU-T SG 7 (now in SG 4) and ISO have started new work (ODMA) that expands the OSI management architecture to allow the distribution of resources being managed. The work applies the ODP concepts mentioned above to management. Efforts are also beginning in TMN to consider extensions of the architecture to allow distributed management.

The requirements considered include: delegation of management activities from one manager to another and the resulting need for coordination of distributed management activities, support for distribution transparency, transparency to different communication protocols, portability of management applications, and guidelines for migration of existing models in a distributed environment.

ODMA is described using the five viewpoints of ODP, with the understanding that the extended architecture will have to accommodate systems based on the traditional peer-to-peer approach defined in X.701|ISO/IEC 10040. A mapping between the terminology used in ODP and OSI Systems management is included in ODMA. For example, "request for an operation or notification" is equivalent to the notion of "invocation." An example is also provided as to how to specify existing OSI management standards in the ODP format. Enterprise viewpoint is used to describe the requirements for a function. ODMA uses Rumbaugh's technique (*The Object Modeling Technique* by J. Rumbaugh) to describe the information without the object interface details. The computational viewpoint is provided by the GDMO and GRM definitions. The interface signatures may be specified using the Interface Definition Language (IDL) from Object Management Group (OMG) or CMIS services. The engineering viewpoint is specified in terms of the communication protocols, specifically using the features available in CMIP.

ODP as well as ODMA standards define concepts such as the viewpoints and apply them to management as a distributed application. However, no notational techniques (except in text) for defining these viewpoints are specified in these standards. ITU-T SG 15 has applied these viewpoints concepts for developing a network-level model for transmission. To define the computational and information viewpoints, a GDMO like notation was introduced.[43]

In order to support distribution in management, ODMA has introduced generic functions (using the viewpoints) and the following computational objects: operations dispatcher, notifications dispatcher, and policy enforcer. In describing the object interactions, ODMA introduces manager role object, which was not present in the existing systems management architecture.

1.34. COMMON OBJECT REQUEST BROKER ARCHITECTURE (CORBA)

The above two sections addressed the concepts introduced for developing distributed application specification in general and management as a specific case. These architectures do not address the technology viewpoint that pertains to

---

[43] *Soap box*: Some have claimed the reason for introducing these notations is to develop a protocol-independent information model. However, it is clear that the schema has been developed with two or three protocols in mind because of some of the restrictions placed on the syntax for communications exchange. While a more rigorous formalism for defining behavior within a schema is very useful, claims that the new notation results in a protocol-independent model have not been proven. A structured text at a higher level than syntax details is possibly a better approach for a protocol-independent information model. Introducing new notations may be a nice academic exercise but may not be palatable to implementors. New compilers will have to be developed to ensure that the syntax is correct and will further delay implementations.

implementation details. One important goal of CORBA is to build portable implementations.

Addressing the implementation issues is an architecture developed by Object Management Group, known as CORBA. Details are available in a series of specifications from X/Open. This architecture provides a flexible approach to integrate a variety of object system implementations, regardless of whether or not these objects are supporting management application. The architecture is built on the software engineering concepts of client and server similar to that described in ODP. The Object Request Broker (ORB) shields from the client the details of the object implementation, the programming language used, and its location. The client may use specific stubs or the interface independent method to invoke operations on an object. The core of the Object Request Broker locates the specific implementation and transfers parameters to a skeleton. An object adapter may be required for the object implementation to request services such as security, and mapping the object reference to implementation. Once the requested operation is completed, the result is returned to the client. The object interfaces within the CORBA architecture defined by OMG uses Interface Definition Language (IDL). As part of CORBA, Application Programming Interfaces are defined to ease the development of distributed applications.

In this chapter, we have either discussed in detail or referenced three different object modeling technologies: models used with CMIP, those with SNMP, and CORBA. These were developed to meet different goals. In order to provide an environment where the strengths of these approaches are combined, a Joint Inter-Domain Management group (JIDM) was formed by X/Open and Network Management Forum (NMF). As part of this effort to provide an interoperable environment, translations of concepts and notations between the three methods are provided. Because of the differing power of the three techniques, reconciling the models for differences is also included. The aim here is to provide much of the mapping that can be automated in a gateway so that interoperability between the varied implementations can be achieved.

It is expected that some of these concepts may be applied when TMN architecture is enhanced in the next study period. Arguments for introducing some of the complexity relative to the benefits offered continue in the standards group; however, the jury is still out on what will succeed in the marketplace. Combining powerful techniques is a goal, and several of the concepts are still maturing. Time will tell the winner.

1.35. SUMMARY

This chapter focuses on information modeling, specifically that used in network management. Several network management systems that exist in the service providers network use a message-based paradigm. The concept of modeling management information, thus providing a rigorous formalism, is an essential part of two-network management methods in the industry. This chapter discusses the various object-oriented information modeling concepts such as inheritance, relationships, and how to model various characteristics of the resource being managed. These concepts were described using examples drawn from existing TMN and Internet management. A glimpse of other modeling efforts and the influence of distributed processing from the latest advances in software development are provided to suggest future directions to the reader.

References

[1] CCITT Recommendation X.720 (1992) | ISO/IEC 10165-1 (1992). Information Technology—Open Systems Interconnection—Structure of Management Information: Management Information Model.

[2] CCITT Recommendation X.721 (1992) | ISO/IEC 10165-2 (1992). Information Technology—Open Systems Interconnection—Structure of Management Information: Generic Management Information.

[3] CCITT Recommendation X.722 (1992) | ISO/IEC 10165-4 (1992). Information Technology—Open Systems Interconnection—Structure of Management Information: Guidelines for the Definition of Managed Objects.

[4] ISO/IEC 18824 (1990). Information Technology—Open Systems Interconnection—Specification of Abstract Notation One (ASN.1) | CCITT Recommendation X.208. (1988). Specification of Abstract Syntax Notation One. Geneva.

[5] CCITT Recommendation G.773 (1990). Protocol Suites for Q-Interfaces for Management of Transmission Systems. Geneva.

[6] CCITT Recommendation G.774. SDH Management Information Model for the Network Element View.

[7] CCITT Recommendation M.3010. Principles for a Telecommunications Management Network (TMN).

[8] CCITT Recommendation M.3020. TMN Interface Specification Methodology.

[9] CCITT Recommendation M.3100. Generic Network Information Model.

[10] CCITT Recommendation M.3180. Catalogue of TMN Managed Objects.

[11] CCITT Recommendation M.3200. TMN Management Services: Overview.

[12] CCITT Recommendation M.3300. F-Interface Management Capabilities.

[13] CCITT Recommendation M.3400. TMN Management Functions.

[14] CCITT Recommendation Q.811. Lower Layer Protocol Profiles for the Q3 Interface.

[15] CCITT Recommendation Q.812. Upper Layer Protocol Profiles for the Q3 Interface.

[16] CCITT Recommendation Q.821. Stage 2 and Stage 3 Description for the Q3 Interface.

[17] CCITT Recommendation X.701 | ISO/IEC 10040: 1992. Information Technology—Open Systems Interconnection—Systems Management Overview.

[18] CCITT Recommendation X.710. (1991). Common Management Information Service Definition for CCITT Applications.

[19] CCITT Recommendation X.711 | ISO/IEC 9596-1: 1991 (E). Information Technology—Open Systems Interconnection—Common Management Information Protocol Specification—Part 1: Specification, Edition 2.

[20] CCITT Recommendation X.712 | ISO/IEC 9596-2: 1992 (E). Information Technology—Open Systems Interconnection—Common Management Information Protocol—Part 2: Protocol Implementation Conformance Statement (PICs) Proforma.

[21] CCITT Recommendation X.723 | ISO/IEC 10165-6: 1992. Information Technology—Open Systems Interconnections—Structure of Management Information—Part 6: Requirements and Guidelines for Implementation Conformance Statement Proformas Associated with Management Information.

[22] CCITT Recommendation X.724 | ISO/IEC 10165-5: Information Technology—Open Systems Interconnections—Structure of Management Information—Part 5: Generic Managed Information, ISO/IEC JTC1/SC21 N6572, February 20, 1992.

[23] CCITT Recommendation X.730 | ISO/IEC 10164-1: Information Technology—Open Systems Interconnections —Systems Management—Part 1: Object Management Function, ISO/IEC JTC1/SC21 N6355, October 15, 1991.

[24] CCITT Recommendation X.731 | ISO/IEC 10164-2: Information Technology—Open Systems Interconnections—Systems Management—Part 2: State Management Function, ISO/IEC JTC1/SC21 N6356, October 15, 1991.

[25] CCITT Recommendation X.732 | ISO/IEC 10164-3: Information Technology—Open Systems Interconnections—Systems Management—Part 3: Attributes for Representing Relationships, ISO/IEC JTC1/SC21 N6357, October 15, 1991.

[26] CCITT Recommendation X.733 (1992) | ISO/IEC 10164-4: 1992. Information Technology—Open Systems Interconnection—Systems Management: Alarm Reporting Function.

[27] CCITT Recommendation X.734 | ISO/IEC 10164-5: 1992. Information Technology—Open Systems Interconnection—Systems Management: Event Report Management Function.

[28] CCITT Recommendation X.735 | ISO/IEC 10164-6: 1992. Information Technology—Open Systems Interconnection—Systems Management: Log Control Function.

[29] CCITT Recommendation X.736 | ISO/IEC 10164-7: Information Technology—Open Systems Interconnection—Systems Management—Part 7: Security Alarm Reporting Function, ISO/IEC JTC1/SC21 N6367, October 15, 1991.

[30] CCITT Recommendation X.740 | ISO/IEC 10164-8: Information Technology—Open Systems Interconnection—Systems Management—Part 8: Security Audit Trail Function ISO/IEC JTC1/SC21 N7039, June 2, 1992.

[31] CCITT Recommendation X.737 | ISO/IEC 10164-14: Information Technology—Open Systems Interconnection—Systems Management—Part 14: Confidence and Diagnostic Test Categories, 1995.

[32] CCITT Recommendation X.738 | ISO/IEC 10164-13: Information Technology—Open Systems Interconnection—Systems Management—Part 13: Summarization Function, 1994.

[33] CCITT Recommendation X.739 | ISO/IEC 10164-11: Information Technology—Open Systems Interconnection—Systems Management—Part 11: Workload Monitoring Function, 1993.

[34] CCITT Recommendation X.741 | ISO/IEC 10164-9: Information Technology—Open Systems Interconnection—Systems Management—Part 9: Objects and Attributes for Access Control, 1995.

[35] CCITT Recommendation X.742 | ISO/IEC 10164-10: Information Technology—Open Systems Interconnection—Systems Management—Part 10: Accounting Meter Function, 1994.

[36] CCITT Recommendation X.745 | ISO/IEC 10164-12: Information Technology—Open Systems Interconnection—Systems Management—Part 12: Test Management Function, 1993.

[37] CCITT Recommendation X.746 | ISO/IEC 10164-15: Information Technology—Open Systems Interconnection—Systems Management—Part 15: Scheduling Function, 1994.

[38] ISO/IEC 7498-4: 1989. Information Processing Systems—Open Systems Interconnection—Basic Reference Model—Part 4: Management Framework.

[39] ISO/IEC ISP 11183-1: Information Technology—International Standardized Profiles—OSI Management—Management Communications Protocols—Part 1: Specification of ACSE, Presentation and Session Protocols for the use by ROSE and CMISE, May 1992.

[40] ISO/IEC ISP 1183-2, Information Technology—International Standardized Profiles—ISO Management—Management Communications Protocols—Part 2: AOM12—Enhanced Management Communications, June 1992.

[41] ISO/IEC ISP 1183-3, Information Technology—International Standardized Profiles—OSI Management—Management Communications Protocols—Part 3: AOM11—Basic Management Communications, May 1992.

[42] ISO/IEC ISP 12059-0, Information Technology—International Standardized Profiles—OSI Management—Common Information for Management Functions—Part 0: Common definitions for management function profiles, 1994.

[43] ISO/IEC ISP 12059-1, Information Technology—International Standardized Profiles—OSI Management—Common Information for Management Functions—Part 1: Object Management, 1994.

[44] ISO/IEC ISP 12059-2, Information Technology—International Standardized Profiles—OSI Management—Common Information for Management Functions—Part 2: State Management, 1994.

[45] ISO/IEC, Information Technology—International Standardized Profiles—OSI Management—Common Information for Management Functions—Part 3: Attributes for Representing Relationships, 1994.

[46] ISO/IEC ISP 12059-4, Information Technology—International Standardized Profiles—OSI Management—Common Information for Management Functions—Part 4: Alarm Reporting, 1994.

[47] ISO/IEC ISP 12059-5, Information Technology—International Standardized Profiles—OSI Management—Common Information for Management Functions—Part 5: Event Report Management, 1994.

[48] ISO/IEC ISP 12059-6, Information Technology—International Standardized Profiles—OSI Management—Common Information for Management Functions—Part 6: Log Control, 1994.

[49] ISO/IEC 12060-1, Information Technology—International Standardized Profiles AOM2n OSI Management—Management Functions—Part 1: AOM211—General Management Capabilities, 1994.

[50] ISO/IEC 12060-2, Information Technology—International Standardized Profiles AOM2n OSI Management—Management Functions—Part 2: AOM212—Alarm Reporting and State Management Capabilities, 1994.

[51] ISO/IEC 12060-3, Information Technology—International Standardized Profiles AOM2n OSI Management—Management Functions—Part 3: AOM213—Alarm Reporting Capabilities, 1994.

[52] ISO/IEC 12060-4, Information Technology—International Standardized Profiles AOM2n OSI Management—Management Functions—Part 4: AOM221—General Event Report Management, 1994.

[53] ISO/IEC 12060-5, Information Technology—International Standardized Profiles AOM2n OSI Management—Management Functions—Part 5: AOM231—General Log Control, 1994.

[54] ISO/IEC TR 10000-1, Information Technology—Framework and Taxonomy of International Standardized Profiles—Part 1: Framework.

[55] Sloman, M. (ed), *Network and Distributed Systems Management*, Addison-Wesley 1994.

[56] Shaler S., and Mellor, S., *Object Lifecycles—Modeling the World in States*, Prentice-Hall, 1992.

[57] Shaler, S., and Mellor, S., *Object-Oriented Systems Analysis—Modeling the World in Data*, Prentice-Hall, 1988.

[58] Flavin, M., *Fundamental Concepts of Information Modeling*, Prentice-Hall, 1981.

[59] Stallings, W., *Networking Standards—A Guide to OSI, ISDN, LAN, and MAN Standards*, Addison-Wesley, 1993.

[60] Draft ITU-T Recommendation X.703|ISO/IEC DIS 13244, Information Technology, Open Distributed Processing Management Architecture (ODMA), 1997.

[61] ISO/IEC 10746-1|ITU Rec. X.901 Information Technology—Open Distributed Processing—Reference Model—Part 1: Overview and Guide to Use, 1995.

[62] ISO/IEC 10746-2|ITU Rec. X.902 Information Technology—Open Distributed Processing—Reference Model: Foundations, 1995.

[63] ISO/IEC 10746-3|ITU Rec. X.902 Information Technology—Open Distributed Processing—Reference Model: Architecture, 1995.

[64] ISO/IEC 10746-4|ITU Rec. X.902 Information Technology—Open Distributed Processing—Reference Model: Architectural Semantics, 1995.
[65] Draft ISO/IEC 13235|ITU Rec. X.9tr Information Technology—ODP Trading Function.
[66] Common Object Request Broker: Architecture and Specification, OMG and X/Open Revision 1.2, 1993.
[67] X/Open Preliminary Specification: Part 2 Object Model Comparison, April 1995.

George Pavlou
Department of Computer Science,
University College London

Chapter 2

OSI Systems Management, Internet SNMP, and ODP/OMG CORBA as Technologies for Telecommunications Network Management

2.1. INTRODUCTION AND OVERVIEW

In this chapter, we compare the ISO/ITU-T OSI Systems Management (OSI-SM) [1], Internet Simple Network Management Protocol (SNMP) [2,3], and ODP/OMG Common Object Request Broker Architecture (CORBA) [5] approaches to network, service, and distributed applications management. The chapter also provides a tutorial overview of those technologies, but it assumes a basic understanding of data network and distributed system principles.

OSI and Internet management have adopted the manager-agent paradigm. Manageable resources are modeled by managed objects at different levels of abstraction. Managed objects encapsulate the underlying resource and offer an abstract access interface at the object boundary. The management aspects of entities such as Network Elements (NEs) and distributed applications are modeled through "clusters" of managed objects, seen collectively across a management *interface*. The latter is defined through the formal specification of the relevant managed object types or classes and the associated access mechanism, that is, the management access service and supporting protocol stack. Management interfaces can be thought of as "exported" by applications in agent roles and "imported" by applications in manager roles. Manager applications access managed objects across interfaces in order to implement management policies. Distribution and discovery aspects are orthogonal to management interactions and are supported by other means. Both OSI and Internet management are primarily communications frameworks. Standardization affects the way in which management information is modeled and carried across systems, leaving deliberately unspecified aspects of their internal structure. The manager-agent model is shown in Figure 2.1. Note that manager and agent applications contain other internal objects that support the implementation of relevant functionality. Since these are not visible externally, they are depicted with dotted lines.

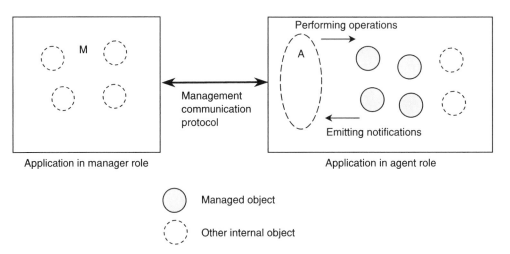

Figure 2.1 The Manager-Agent Model.

The manager and agent roles are not fixed, and management applications may act in both roles. This is the case in hierarchical management architectures such as the Telecommunications Management Network (TMN) [6]. In hierarchical management models, managed objects also exist in management applications, offering views of the managed network, services, and applications at higher levels of abstraction. Management functionality may be organized in different layers of management responsibility: element, network, service, and business management according to the TMN model. Management applications may act in dual manager-agent roles, in either peer-to-peer or hierarchical relationships. Figure 2.2 shows three types of management organization according to the manager-agent model: centralized, flat, and hierarchical. The centralized model is best exemplified by SNMPv1 Management Operation Centers (MOCs). The flat model reflects the evolution of SNMPv1 to SNMPv2, with "manager-to-manager" capabilities. Finally, the hierarchical model is best exemplified by TMN, which uses OSI management as its base technology. Note that in both flat and hierarchical models, management applications are hybrid, assuming both manager and agent roles.

Although OSI and Internet management conform to the same broad model, within their constraints there is room for wide divergence, and, to an extent, they are at two ends of the spectrum. Internet management was designed to address the management needs of private data networks (LANs/MANs) and Internet backbones. As such, it has adopted a connectionless (CL), polling-based, simple

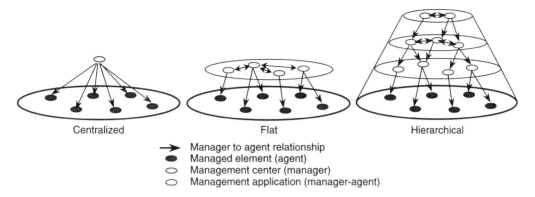

Figure 2.2 Models of Management Organization.

"remote-debugging" approach, opting for agent simplicity and projecting a centralized or flat organizational structure. On the other hand, OSI management was designed to address the management needs of OSI data networks and telecommunications environments. As such, it has adopted a connection-oriented (CO), event-driven, fully object-oriented (O-O) approach, opting for generality and trying to move management intelligence close to managed elements. It has been adopted as the base technology for TMN, and it supports hierarchical organizational structures.

ISO/ITU-T Open Distributed Processing (ODP) [4] is a general framework for specifying and building distributed systems. The Object Management Group (OMG) Common Object Request Broker Architecture [CORBA] can be seen as its pragmatic counterpart. While SNMP and OSI management are *communications* frameworks, ODP/OMG CORBA target a *programmatic* interface between objects in client or server roles and the underlying support environment, that is, the Object Request Broker (ORB). Server objects are accessed through interfaces on which operations are invoked by client objects in a location-transparent fashion. Choices made by the Internet Engineering Task Force (IETF) and ISO/ITU-T on the one side and OMG on the other side reflect their different preoccupations: management communications for the former and distributed software systems for the latter. The difference in approach is sometimes referred to as "vertical" versus "horizontal" interfaces. Vertical interfaces standardize communications interactions between systems. The horizontal approach standardizes Application Programming Interfaces (APIs) which are used to "plug" application objects on the global supporting infrastructure. The latter is also referred to as the Distributed Processing Environment (DPE) and encapsulates the underlying network, hiding heterogeneity and providing various transparencies. The ODP/OMG CORBA model is shown in Figure 2.3.

The OMG CORBA paradigm is that of a client-server, with distribution provided through the ORB. The unit of distribution is the single object as opposed to the OSI and Internet object cluster. Client and server CORBA objects communicate through the ORB, whose services are accessed through standard APIs. Interoperability is achieved through the formal specification of server interfaces, the ORB APIs, and the underlying inter-ORB protocols. One key difference to

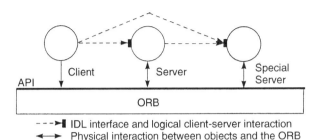

IDL interface and logical client-server interaction
Physical interaction between objects and the ORB

Figure 2.3 The OMG CORBA Model.

OSI and Internet management is that the object model and APIs have been addressed first, while the underlying protocols may be replaced. Of course, interoperability dictates an agreed protocol, but the rest of the framework is not heavily dependent on it. The key benefit is portability of objects across different CORBA implementations due to the standard ORB APIs and the various transparencies that are (or will be) supported by the latter. Note that communication aspects are "hidden" inside the ORB and, as such, are not shown in Figure 2.3. While OMG CORBA is a general distributed systems framework, its object-oriented nature and the fact that management systems are composed of interacting objects suggest that it could also be used for management. OMG CORBA has been chosen by the Telecommunication Information Network Architecture (TINA) [7] initiative as the basis for their DPE. TINA aims at a framework for future advanced services, including service and network management aspects.

In this chapter, we examine aspects of those technologies, assessing their suitability in the context of telecommunications management. In section 2.2, we examine information modeling aspects in the three frameworks and assess their flexibility and expressiveness. In section 2.3, we examine the access and distribution aspects, including the underlying communications and the paradigm of operation. In section 2.4, we examine other aspects, including applicability in various contexts, generic functionality, and security. In section 2.5, we examine issues addressing their interworking and coexistence. Finally, we present a summary and discuss potential future directions in telecommunications management.

2.1.1. Background Information: ASN.1

Some necessary background information concerns the OSI Abstract Syntax Notation 1 (ASN.1) [8] language. This is an abstract "network" data structuring language that supports simple and constructed types. A particularly important ASN.1 type is the Object Identifier (OID), which expresses a sequence of nonnegative integers on a global registration tree. OIDs are registered by standards bodies, for example, ISO, ITU-T, and IETF, and are used instead of friendly string names to avoid ambiguities. For example, the OSI *objectClass* attribute name is registered as {joint-iso-ccitt(2) ms(9) smi(3) part2(2) attribute(7) objectClass(65)}. ASN.1 is used in both the OSI management and SNMP frameworks to specify the management protocol packets and the structure of managed object information, for example, attributes, operation parameters and results, and notification information.

2.2. MANAGEMENT INFORMATION MODELS

A management framework and associated technology should be applicable to network, service, system, and distributed application management. In addition, the applications and support infrastructure that constitute the management system should also be manageable. The ideal information model must cope easily with management information related to such a variety of management targets. At the same time, it must impose a measure of uniformity on the structure of management information so that it is possible to devise a set of generic management operations that are applicable in all management contexts. The original motivation for developing the three frameworks was different: network management for SNMP; network and service management for OSI; and distributed application operation and management for OMG CORBA. As a result, the relevant information models, despite exhibiting similarities, have important differences.

A key difference between the SNMP and the OSI Management/OMG CORBA information models is that the latter have enthusiastically endorsed the O-O approach and have made full use of relevant concepts such as classes and inheritance. In the case of SNMP, the information model is often referred to as Object-based, with classes and inheritance regarded as unnecessary complications and thus deemed undesirable. All three models are quite general and, despite their differences, anything that can be modeled in one can also be modeled in another. In fact, methodologies have been developed for converting between them as described in section 2.5. The key issue is whether the greater expressive power one offers results in the better abstraction of management information and is worth the price of the additional complexity. Figure 2.4 shows a pictorial view of the notion of objects in

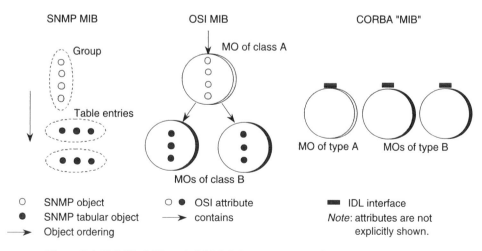

Figure 2.4 SNMP, OSI, and CORBA Management Information Base.

the three frameworks and their collective view as a Management Information Base (MIB) across a management interface. Information modeling aspects are explained next, in a separate subsection for each framework.

2.2.1. The SNMP Information Model

SNMP information modeling principles are collectively referred to as the Structure of Management Information (SMI) and are specified in [9] for SNMPv1 and in [10] for SNMPv2, the latter being an extension of the SNMPv1 model.

The basic building block of an SNMP MIB is the *object*. Objects belong to a particular *object type* and have *values*. According to the SNMP SMI, object values are restricted to a very small set of allowed syntaxes, resulting ultimately in the basic ASN.1 types INTEGER, OCTET STRING, and OBJECT IDENTIFIER. Other *application-wide* types such as Counter, Gauge, NetworkAddress, and Timeticks must resolve to either integer or string scalar types. The only constructed type allowed is a simple two-dimensional table consisting of elements of the previous primitive types. Great emphasis is given to the fact that the allowable syntaxes are few, simple, and scalar. The key advantage claimed for this approach is that object values carried across a network must be encoded in a "network-standard" way. Encodings and decodings can be computationally expensive, especially if the syntaxes involved are complex. Because SNMP uses only a small fixed set of syntaxes, it is possible to hand-code the encoding and decoding software in an optimal fashion. Thus, the SNMP SMI is analogous to a computer language that has a small set of basic types together with two-dimensional arrays.

The SNMP SMI defines a notation for specifying the properties of new object types, an ASN.1 *macro* called OBJECT-TYPE, while ASN.1 is used to specify the object syntaxes and the tabular structure. SNMP object types can be either single or multiple instanced, with multiple-instanced objects allowed only in tables. SNMP objects are similar to OSI Management/OMG CORBA attributes, while there is no notion of a "composite" object-boundary that encapsulates a number of scalar objects modeling a manageable entity. The only composite relationship relates objects to tables.

A table has rows (also referred to as table entries or records), with each row represented by a *SEQUENCE* ASN.1 type that contains a statically defined number of objects. The table itself is modeled as *SEQUENCE OF* ASN.1 type with respect to the rows or entries, allowing an arbitrary number of those to be dynamically instantiated. A table thus resembles a "dynamic array of records." Tables may grow to arbitrary length but must always be of fixed width. A further rule restricts the syntaxes used within a row to be "scalar"; thus, one cannot define tables within tables. Note also that tables and table entries are only conceptual composite objects in SNMP: only the individual scalar objects that constitute a table entry are accessible through the management protocol.

Let us examine the use of this modeling framework through an example. A typical single-instanced set of objects are, for example, those modeling aspects of a connection-oriented transport protocol entity such as ISO Transport Protocol (TP) and Internet Transmission Control Protocol (TCP). Such objects will cover the number of current and previous connections, the number of incoming and outgoing unsuccessful connection requests, the number of transport packets sent, received, and retransmitted, and the number of various protocol-related errors. All these will have to be separate objects, loosely related through a "transport protocol" group. Note that this model does not support multiple instances of a transport protocol per node. If the latter was necessary, a table of transport protocol entries would be needed, but we will overlook this restriction for simplicity. The group will also comprise transport connections, which are multiple-instanced objects and have to be modeled through a table. A tpConnTable may be defined as a SEQUENCE OF tpConnEntry, with tpConnEntry being a SEQUENCE of objects modeling various aspects of the connection, such as source and destination access points, the connection state, and so on. Figure 2.4 shows the objects of a single-instanced group, for example, tp group, and two table entries, for example, tpConnEntry. Note that both the group and table entries are depicted with dotted lines since there is no notion of a composite object boundary in the SNMP information model.

When objects are instantiated, they must be named so that they can be addressed unambiguously. The SNMP framework uses object identifiers for naming. Every *object type* has a registration OID, while an *object instance* is identified by the object type OID, suffixed by a part that uniquely identifies that instance. For non-tabular objects any suffix would do, so the minimal *.0* is used. For example, *tpCurrentConnections.0* is the instance of the object type *tpCurrentConnections*. In the case of multiple-instanced tabular objects such as the *tpConnState of the tpConnEntry,* the suffix needs to signify the table entry. The latter can be constructed from the values of one or more objects of that entry that constitute the "key," as specified in the relevant OBJECT-TYPE template for that table entry. In our example, the values of the *tpConnSrcAddr* and *tpConnDestAddr* objects may be used as they uniquely identify each connection. For example, *tpConnState.123.456* is the instance of the object type *tpConnState*, corresponding to the connection with source address 123 and destination address 456. Note that when object values are strings as opposed to integers, they need to be converted to object identifier suffixes: the SNMP SMI specifies rules for this conversion.

The SNMP naming architecture exhibits a very tight coupling between object type and instance identifiers. The problem with it is that instances of a particular object type can only appear in one particular place in the registration tree. This means that one cannot define generic object types to be used in several contexts. For example, a common object for representing the state as perceived by a managed resource is the *operationalState*. In the SNMP framework, it is not possible to define such a generic object, but specific objects have to be defined for every particular context, for example, *tpOperationalState*. An additional problem is that because this is not the intended use of OIDs, it has been found that most SNMP implementations spend a lot of processing time scanning through object names, trying to separate type and instance information. Finally, object names formed through OID suffixes are not natural.

SNMP objects accept only *Get* and *Set* operations. The Set operation can be performed only to objects that have a *read-write* or *read-create* access level according to the object specification. Access to objects with Get and Set operations is subject to the access control policy across a management interface. SNMP objects support neither *Create* nor *Delete* operations explicitly. Create and Delete semantics are implicitly supported, however, for multiple-instanced objects, that is, table entries or rows. In SNMPv1, row creation may be requested through a Set operation for an entry that is currently not in the table by setting the values of all the row objects. This type of Set operation is interpreted as table entry creation; however, this behavior is not mandated and implementations may behave differently. In addition, the SNMP protocol limits the maximum Protocol Data Unit (PDU) or packet size, so it might not be possible to pass values for all the row objects! SNMPv2 remedies this defect and uses an elaborate interaction scheme to ensure atomicity of row creation. This is supported through the use of a *RowStatus* object which must be present in any table that allows row creation. Creation can follow either a *createAndGo* or *createAndWait* protocol, according to the value of rowStatus. The former is similar to the SNMPv1 creation style and restricted by the maximum PDU size. In the latter, the values of the rowStatus object progress successively through *notReady* -> *notInService* -> *active*, while the row is being created through more than one Set requests. Row deletion is achieved through a Set operation that sets the value of row status to *destroy*. Although all this works, it is certainly not simple. In fact, a high price has been paid in order to avoid introducing Create and Delete object operations through separate protocol primitives.

In both the OSI Management and OMG CORBA models, objects may accept arbitrary actions that operate on the object boundary, as described next. Given the fact that SNMP objects are essentially attributes compared to OSI and CORBA objects, imperative actions with arguments and results are meaningless. Nevertheless, actions are necessary and may be modeled in SNMP by objects that support the arguments and results of an action. An "action" may be emulated by a Set request, followed possibly by a Get request to retrieve the results. For example, a "reboot" action may be modeled by a boolean *rebootState* object whose value is set to *true*. (In this case, there is no action result.) This type of emulation is not very elegant and may result in complex interactions and an awkward object model for imperative commands with complex "argument and result" parameters.

In SNMPv1, agent applications may emit notifications, called *traps*, associated with the SNMP protocol rather than a MIB specification. These are supposed to be used only for a small and predefined set of events (*warmStart, coldStart, linkUp, linkDown, authenticationFailure*). However, some MIB designers (notably those of the Remote Monitoring MIB [11]) have extended the concept and have provided notations that allow MIB designers to specify resource-specific notifications and the information that should be included in them. A notation of this type has now been included in the SNMPv2 SMI [10].

2.2.2. The OSI System Management Information Model

Chapter 1 discussed OSI-SM and TMN information modeling in detail. The OSI-SM Information Model (MIM) is defined in [12]. An OSI Management Information Base (MIB) defines a set of *Managed Object Classes* (MOCs) and a schema that defines the possible containment relationships between instances of those classes. There may be many types of relationships between classes and their instances, but containment is treated as a primary relationship and is used to yield unique names. The smallest reusable entity of management specification is not the object class, as is the case in other O-O frameworks, but the *package*. Object classes are characterized by one or more mandatory packages while they may also comprise conditional ones. An instance of a class must always contain the mandatory packages while it may or may not contain conditional ones. The latter depends on conditions defined in the class specification. Managing functions may request that particular conditional packages are present when they create a managed object instance.

A package is a collection of attributes, actions, notifications, and associated behavior. Attributes are analogous to object types in SNMP. Like an object type, an attribute has an associated syntax. Unlike SNMP, however, there are no restrictions on the syntax used. If desired, an attribute need not be of scalar type. A number of useful generic attribute types have been defined in [13], namely, *counter*, *gauge*, *threshold*, and *tide-mark*. A MIB designer may derive resource-specific types from these. Support for arbitrary syntaxes provides a much more flexible scheme than that of SNMP. For example, it allows the definition of complex attributes such as a *threshold,* whose syntax can include fields to indicate whether or not the threshold is currently active and its current value. It is also more expensive to implement since support for encoding and decoding of completely arbitrary syntaxes must be provided.

OSI managed object classes and packages may have associated specific actions that accept arguments and return results. Arbitrary ASN.1 syntaxes may be used, providing a fully flexible "remote method" execution paradigm. Exceptions with MOC-defined error information may be emitted as a result of an action. The same is also possible as a result of operations to attributes under conditions that signify an error, for which special information should be generated. Object classes and packages may also have associated notifications, specifying the condition under which they are emitted and their syntax. The latter may again be an arbitrary ASN.1 type. By behavior, one means the semantics of classes, packages, attributes, actions, and notifications and the way they relate, as well as their relationship to the entity modeled through the class. Behavior is specified in natural language only; the same is true in SNMP.

OSI Management follows a fully O-O paradigm and makes use of concepts such as inheritance. Managed object classes may be specialized through subclasses that inherit and extend the characteristics of superclasses. This allows reusability and extensibility of both specification and associated implementation if an object-oriented design and development methodology is used. Pursuing the previous example, a *tpProtocolEntity* object class may inherit from an abstract *protocolEntity* class that models generic properties of protocol entities, such as the operational state and the service access point through which services can be accessed. By abstract class is meant a class that is never instantiated as such but serves only inheritance purposes. In the same fashion, an abstract *connection* class may model generic properties of connection-type entities such as the local and remote service access points, the connection state, and creation and deletion notifications. The inheritance hierarchy of those classes is shown in Figure 2.5. It

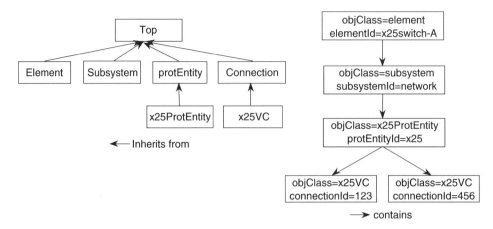

example name: {subsystemId=network, protEntityId=x25, connectionId=123}

Figure 2.5 Example OSI Inheritance and Containment Hierarchies.

should be noted that conditional packages allow for dynamic (i.e., run-time) specialization of an object instance, while inheritance allows only for static (i.e., compile-time) specialization through new classes.

The specification of manageable entities through generic classes that are used only for inheritance and reusability purposes may also result in generic managing functions by using *polymorphism* across management interfaces. For example, it is possible to provide a generic connection-monitor application that is developed with the knowledge of the generic *connection* class. This may monitor connections in different contexts, for example, X.25 and ATM, disregarding the specialization of a particular context. That way, reusability is extended to managing functions as well as managed object classes and their implementations.

In OSI management, a derived class may extend a parent class through the addition of new attributes, actions, and notifications; through the extension or restriction of the value ranges; and through the addition of arguments to actions and notifications. Multiple inheritance is also allowed, and it has been used extensively by information model designers in standards bodies. Despite the elegant modeling that is possible through multiple inheritance, such models cannot be easily mapped onto O-O programming environments that do not offer such support, for example, Smalltalk and Java. Multiple inheritance is a powerful O-O specification technique but increases system complexity.

A particularly important aspect behind the use of object-oriented specification principles in OSI management is that they may result in the allomorphic behavior of object instances. *Allomorphism* is similar to polymorphism but has the inverse effect: in polymorphism, a managing function knows the semantics of a parent class in an inheritance branch and performs an operation to an instance that responds as the leaf class. In allomorphism, that instance should respond as the parent class, hiding completely the fact that it belongs to the leaf class. For example, a polymorphic connection monitor application can be programmed to know the semantics of the *connection* class and only the syntax of specific derived classes through meta-data. When it sends a "read all the attributes" message to a specific connection object instance, for example, x25Vc, atmVcc, it wants to retrieve all the attributes of that instance, despite the fact that it does not understand the semantics of the specific "leaf" attributes. In allomorphism, a managing function programmed to know a *x25ProtocolEntity* class should be able to manage instances of a derived *x25ProtocolEntity2* class without knowing of this extension at all. In this case, operations should be performed to the *x25ProtocolEntity2* instance as if it were an instance of the parent *x25ProtocolEntity* class, since the derived class may have changed the ranges of values, added new attributes, arguments to actions, and so on.

Polymorphism is a property automatically supported by O-O programming environments, while allomorphism is not and has to be explicitly supported by management infrastructures. Allomorphic behavior may be enforced by sending a message to an object instance and passing to it the object class as an additional parameter, essentially requesting the object to behave as if it were an instance of that class. When no class is made available at the object boundary, the instance behaves as the *actual* class, that is, the leaf class in the inheritance branch. Allomorphic behavior is considered very important since it allows the controlled migration of management systems to newer versions by extensions of the relevant object models through inheritance, while still maintaining compatibility with the past. This is particularly important in management environments, for requirements and understanding of the problem space are expected to be continuously evolving. Finally, it should be mentioned that allomorphism hides extensions at the *agent* end of the manager-agent model. Extensions in managing systems should be hidden by programming them to revert to the "base" information model if this is what it is supported across a management interface. Though possible, this requires additional effort and increases complexity.

The root of the OSI inheritance hierarchy is the *top* class which contains attributes self-describing an object instance. These attributes are the *objectClass* whose value is the actual or leaf-most class; *packages* that contain the list of the conditional packages present in that instance; *allomorphs* that contain a list of classes the instance may behave as; and *nameBinding* which shows where this instance is in the naming tree as explained next. For example, in the instance of the x25ProtocolEntity2 class mentioned earlier, objectClass would have the value *x25ProtocolEntity2* and allomorphs would have the value *{x25ProtocolEntity}*. When an instance is created by a managing function, the conditional packages may be requested to be present by initializing accordingly the value of the packages attribute, which has "set by create" properties.

Managed object classes and all their aspects such as packages, attributes, actions, notifications, exception parameters, and behavior are formally specified in a notation known as Guidelines for the Definition of Managed Objects (GDMO) [14]. GDMO is a formal object-oriented information specification language that consists of a set of *templates*. A "piecemeal" approach is followed, with separate templates used for the different aspects of an object class, that is, class, package, attribute, action, notification, parameter, and behavior templates. GDMO formally specifies only syntactic aspects of managed object classes. Semantic aspects, that is, the contents of behavior templates, are expressed in natural language. The use of formal specification techniques such as System Definition Language and Z are considered by the ITU-T in order to reduce the possibility of ambiguities and misinterpretations and to increase the degree of code automation.

The types of attributes, action and notification arguments, replies, and exception parameters are specified as ASN.1 types. Object Identifiers are associated with classes, packages, attributes, notifications, and actions, but they have nothing to do with instance naming. Instead, managed object instances are named through a mechanism borrowed from the OSI Directory [17]. Managed object classes have many relationships, but containment is treated as a primary relationship to yield unique names. Instances of managed object classes can be thought as logically containing other instances. As such, the full set of managed object instances available across a management interface are organized in a Management Information Tree (MIT), also referred to as the "containment hierarchy." This requires that an attribute of each instance serves as the "naming attribute." The attribute and its value form a Relative Distinguished Name (RDN), for example, connectionId = 123. This should be unique for all the object instances at the first level below a containing instance. If these instances belong to the same class, then it is the value of the naming attribute that distinguishes them (the "key").

The containment schema is defined by *name-binding* GDMO templates which specify the allowable classes in a superior/subordinate relationship and identify the naming attribute. Name bindings and naming attributes are typically defined for classes in the first level of the inheritance hierarchy, immediately under *top* so that they are "inherited" by specific derived classes. An example of a containment tree is shown in Figure 2.5, modeling connections contained by protocol entities, by layer subsystems, and by a network element. A managed object name, also known as a Local Distinguished Name (LDN), consists of the sequence of all the relative names from the top of the tree down to the object, for example, *{subsystemId = network, protocolEntityId = x25, connectionId = 123}*. OSI management names are assigned to objects at creation time and last for the lifetime of the object. An OSI managed object has exactly one name; that is, the naming architecture does not allow for multiple names. The same is true for SNMP objects as described, although the naming architecture is different.

While SNMP was likened to a computer language with a few simple types plus arrays, OSI Management can be likened to a full object-oriented language since it allows new types (of arbitrary complexity) to be defined and arbitrary methods (actions) to be invoked upon them.

2.2.3. The ODP/OMG CORBA Information Model

While both SNMP and OSI management are *communications* frameworks, standardizing management interfaces for applications in agent roles, OMG CORBA targets a *programmatic* interface between objects in client or server roles and the underlying support environment, that is, the ORB. Server objects are accessed through interfaces on which operations are invoked by client objects.

The ODP/OMG CORBA information model is fully object-oriented, in a similar fashion to that of OSI management. Objects are characterized by the *interfaces* they support. An ODP object may support multiple interfaces bound to a common state, unlike OSI management where objects may have only one interface. The current OMG specification, however, allows only a single interface per object. In fact, the OMG model defines objects through the specification of the relevant interfaces. As such, there is no direct concept of an object class in OMG. Object interfaces may be specialized through inheritance, while multiple inheritance is also allowed. The root interface in the inheritance hierarchy is of type *Object*. OMG interfaces are specified using the Interface Definition Language (IDL) [15]. The IDL specification technique is more monolithic than the GDMO piecemeal approach: the minimum reusable specification entity is the interface definition as opposed to the individual package, attribute, action, and notification in GDMO. IDL may be regarded as broadly equivalent to the GDMO/ASN.1 combination in OSI management, though less powerful and with some differences highlighted below.

An OMG object may have attributes, accept operations at the object boundary, and exhibit behavior. Such an object is used to implement a computational construct. In a management context, an object may behave as a manageable entity, modeling an underlying resource. Object attributes have associated syntax, which in IDL is called a *type*. Arbitrary syntaxes are allowed, although the expressive power of IDL types is less than ASN.1. There is no mechanism for grouping attributes together as in OSI management. Attributes accept Get and Set operations, while only standard exceptions may signify an error during such operations. This is in contrast to GDMO, where arbitrary class-specific errors and associated information may be defined to model exceptions triggered by attribute-oriented operations. OMG objects also accept object-oriented operations, similar to the GDMO actions. The normal execution of an operation results in a reply, while object-specific exceptions may be defined. Operation invocations, replies, and exceptions may take arbitrary parameters in terms of IDL types. It should be mentioned that a GDMO action may result in multiple replies, despite the fact that information model designers seldom use this feature. Multiple results are not explicitly supported in IDL but may be modeled through "callback" invocations.

A key difference between GDMO and OMG objects is that the latter do not allow for the late binding of functionality to interfaces through optional constructs similar to the GDMO conditional packages. An OMG object type is an absolute indication of the characteristics of an instance of that type. However, attribute and operation parameter values may be "null," while CORBA supports a standard *not_implemented* exception. An additional major difference is that in IDL it is not possible to specify event types *generated* by an object: events are modeled as "operations in the opposite direction." As such, events are specified through operations on the interface of the receiving object. An OMG managed object needs to specify a separate interface containing all the events it can generate; the latter needs to be supported by managing objects that want to receive these events. There are more differences with respect to the way events are disseminated, but these are discussed in section 2.3.

ODP/OMG do not provide a built-in operation for instantiation of interfaces by client or managing objects. The reason for that is that OMG takes a "programmatic" view of object interfaces, and, as such, a create operation is meaningless before that interface exists! While GDMO objects appear to accept create operations according to the specification, the latter are essentially targeted to the agent infrastructure in engineering terms. As such, interface creation in OMG may only be supported by existing interfaces: *factory* objects may be defined that allow client objects to create application specific interfaces. This approach is not flexible, for a factory interface is necessary for every other interface that can be dynamically created. A more generic *factory service* would be welcome, allowing flexibility in the placement of new objects as currently factory objects may place new objects in the same node.

Deletion of objects is possible through the OMG Object Life-Cycle Services [16]. The latter has specified an interface that provides a *delete* as well as *copy/move* operations. Any other interface that needs to be deleted should inherit from the life-cycle interface. The copy/move operations apply to object implementations and appear to be very powerful as they support relocation and replication. The downside is that it is not at all clear how these will be provided. In the absence of implementations supporting life-cycle services at present, interface deletion is currently tackled through the definition of interface-specific *delete* operations. The problem is that if an object receives a delete request through its interface and deletes itself, there can be no reply to the performing client. An exception instead is raised, and the client will never know if deletion was completed successfully or something else went wrong while the object was cleaning up its state. Hopefully, mature implementations of the life-cycle service interface will solve such problems in the future.

In summary, creation and deletion of interfaces is not handled in a fully satisfactory fashion. The main problem is that such facilities should not be seen as separate "services" but should be an integral part of the underlying platform, that is, the ORB. Unfortunately, the OMG did not take this approach. Finally, it should be mentioned that object creation and deletion in distributed system contexts are used mostly for system instantiation and termination, that is, not very frequently while this is *not* the case in management environments.

While the OMG IDL object model has many similarities to GDMO, a marked difference concerns naming. OMG objects can be identified and accessed through *Object References*. The latter are assigned to objects at creation time and are opaque types, that is, have no internal structure and, as such, do not reveal any information about the object. Their typical implementation is through long bit-strings in order to facilitate processing; they are in fact similar to pointers in programming languages. An object may have more than one reference while objects may also be assigned names. The latter are distinct from objects, unlike Internet and OSI management where an object always has a name. Actually, OMG objects need not have names at all, for they may be "bound to" by type through the ORB and accessed through their interface reference(s). In addition, names may be assigned to objects, but this mapping may change at any time. Names are assigned to objects through the Name

Service [16], which provides a directed graph of naming contexts with potentially many roots. A point in the graph may be reached via many routes, which means that an object may have many names. This is in contrast to OSI management where there is a naming tree instead of a naming graph and objects have exactly one name. The name server may be essentially used to assign names to objects and to resolve names to object references.

The example presented previously may be expressed in terms of CORBA objects in a one-to-one mapping with the equivalent OSI-managed objects. A key difference is that there is no need for a containment tree as such, but containment may be treated as any other relationship. Despite that, it will probably be necessary to model containment in order to assign unique names to managed objects, in a similar fashion to OSI management. Those objects may not be "discovered" and selected based on containment relationships, as is the case in OSI management, but through the trader. Discovery and access aspects are addressed in the next section.

Finally, while polymorphism is a general property of object-oriented systems and, as such, is supported in CORBA, there is no notion of allomorphism. The latter may be supported by passing the interface type explicitly as an argument to operations. In this case, though, allomorphism will not be transparent as it is in OSI-SM. In addition, there is no built-in support for the discovery of the allomorphic interface types that an object supports through a facility similar to the OSI-SM *allomorphs* attribute.

2.3. ACCESS AND DISTRIBUTION PARADIGM

In the three management frameworks, managing functions or objects implement management policies by accessing managed objects. By access paradigm, we mean the access and communication aspects between managing and managed objects. Access aspects include both the remote execution of operations on managed objects and the dissemination of notifications emitted by them. Given the different origins of OSI/Internet management and ODP/OMG CORBA, that is, communications and distributed software systems respectively, there are marked differences in the relevant access paradigms. OSI and Internet management follow a protocol-based approach, with message-passing protocols modeling operations on managed objects across a management interface. The operations and parameters supported by those protocols are a superset of those available at the managed object boundary, with the additional features supporting managed object discovery and multiple object access. The protocol operations are addressed essentially to the agent administering the managed objects, which acts as a naming, discovery, access, and notification server. On the other hand, OMG CORBA specifies the API to the ORB through which client objects may perform operations on server objects. Remote operations are supported by a Remote Procedure Call (RPC) protocol. The latter carries the remote operation parameters and results, while functions such as object discovery and multiple object access are left to application services such as naming and trading.

While both OSI and Internet management have tried to optimize access aspects with respect to the target management environments, they have paid less attention to distribution aspects. By distribution, we mean the way in which managing and managed systems and objects discover each other and how various related transparencies, such as location, are supported. In OSI management, distribution has been recently addressed through discovery and shared management knowledge services [18], supported by the OSI Directory [17]. In the SNMP world, distribution is partly addressed through predefined addresses. On the other hand, OMG CORBA is influenced by ODP [4], and, as such, from the beginning it has been designed with distribution and various transparencies in mind. Its ORB-based architecture has targeted the optimal provision of distribution, in the same fashion that the manager-agent architecture adopted by OSI and Internet management has targeted the optimal support for managed object access services. In this section, we look at and compare the access and distribution aspects of the three frameworks.

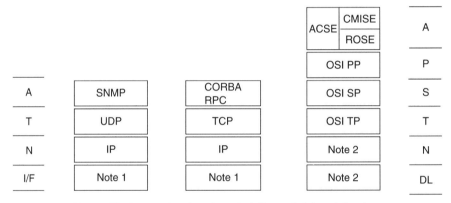

Note 1: The Internet interface layer is deliberately left undefined.
Note 2: There exist many OSI Network/DataLink protocol combinations.

Figure 2.6 SNMP, CORBA, and OSI Protocol Stacks.

2.3.1. The Internet SNMP

In the Internet management framework, the fundamental axiom has been simplicity at the agent part of the manager-agent spectrum. This important design decision aimed at the provision of standard management interfaces to the majority of network devices at a minimal cost and has influenced the associated access paradigm. SNMP has been designed to support a connectionless style of management communication, and, as such, it has been mapped over the User Datagram Protocol (UDP) and the Internet Protocol (IP) as shown in Figure 2.6 [2,3]. The relevant thinking is that reliable transport protocols such as the Transport Control Protocol (TCP) impose too much memory and processing overhead to be supported by simple network devices such as routers and bridges. In addition, maintaining transport connections requires state information, which again is considered to be expensive and, as such, undesirable. Also, bearing in mind that SNMP projects a largely centralized model, it is impossible to maintain simultaneously thousands of connections from a Management Operations Center (MOC) to the managed devices.

Applications using the SNMP protocol will need to ensure the reliability of underlying communication by undertaking retransmission. In other words, applications should try to emulate the functionality of a reliable transport protocol by setting timers every time they perform a transaction and possibly retransmit. An important related aspect is that only applications in manager roles do retransmit; agents simply respond to requests. As such, there is the possibility that either the request or the response packet is lost. If the response packet is lost, the management operation will eventually be performed more than once; this is fine for information retrievals but might cause problems for intrusive operations. As such, either the latter should be *idempotent*, or measures should be taken in the agent to prevent an operation from being performed twice (test-and-set, etc.). Finally, agents in managed devices do not retransmit notifications (*traps*). It is only applications in dual manager-agent roles that are allowed to retransmit notifications (*inform-requests*) in SNMPv2.

The management protocol operations are a superset of the operations available at the managed object boundary. We have already mentioned that SNMP objects accept only Get and Set operations and that imperative commands (actions) and table entry creation and deletion are emulated through Set. As such, the SNMP protocol also has *Get* and *Set* request packets. Agents emit notifications, which are sent to manager applications through the *Trap* SNMP packet. In addition to those operations, manager applications must be able to "discover" transient objects—for example, table entries for which it is impossible to know their names in advance (e.g., connections, routes). We have already mentioned that the SNMP naming architecture relies on ASN.1 object identifiers. Since the latter have ordering qualities, a collection of SNMP objects visible across an interface is ordered in a linear fashion, which means there is a first and a last object. This linear structure was depicted in Figure 2.4. Note that table entries are ordered on a column-by-column basis, which is why they are depicted horizontally in that figure. It is exactly this linear structure that is exploited through the *Get-next* operation, which allows retrieval of the next object or any other object in the MIB. This is typically used for traversing tables and allows the retrieval of one table entry at a time if the entry names are not known in advance.

In SNMPv2, two more primitives have been added. The *inform-request* packet is to be used between hybrid manager-agent applications in order to report asynchronous notifications to each other; its receipt should be acknowledged by the receiving application. This means that it should be retransmitted, so that it can be thought as a "reliable trap." Note, however, that managed elements are not supposed to use this facility. The second addition has been the *Get-bulk* primitive, which is an extension of *Get-next* as it allows retrieval of more than one next objects. Although it is better than *Get-next*, it is still a poor bulk data retrieval facility because SNMP does not allow multiple replies, but the result should fit in one response packet. Given the fact that the underlying transport is unreliable, the maximum allowed application-level packet size is about 500 bytes in order to avoid network-level segmentation. In short, the SNMP mode of operation is request-reply or request only for traps. The possible interactions using SNMP operations are shown in Figure 2.7.

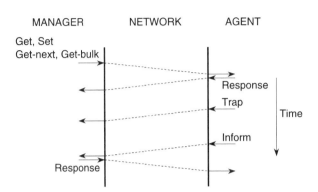

Note 1: Inform is only allowed for dual agent-manager entities.
Note 2: Get-bulk and Inform have been added in SNMPv2.   **Figure 2.7** SNMP Protocol Interactions.

Event-based operation in SNMP is pretty simple-minded. Traps are sent asynchronously from managed devices to predefined managers and are not retransmitted. This means they are inherently unreliable; as such, managers should not rely on them but should also monitor managed devices through periodic polling. This approach does not scale as it imposes management traffic on the managed network even when nothing happens. It also requires careful tradeoff between the polling frequency and the potential elapsed time after a significant change before a manager knows about it. In SNMPv2, the event model between hybrid manager-agent applications is more flexible, for it allows the request of particular events by creating entries in a relevant table. These events are retransmitted until an acknowledgment is received. On the other hand, it is not possible to specify any other event-associated conditions through filtering, in order to minimize further management traffic.

Finally, distribution in terms of location transparency is not an issue in SNMP. Manager applications address agents in managed devices through their IP address, while the agent is always attached to the UDP port number 161. This simplistic model allows for the discovery of device agents in LANs and MANs because of their broadcast nature: a multicast SNMP message is sent to all the nodes of the LAN at the known port number, requesting the first MIB object, for example, *Get-next(name = 0.0)*, and the nodes that respond have been "discovered." On the other hand, such discovery is not possible for higher-level hybrid manager-agent applications, for it is not specified on which port they should be attached. The only aspect that is specified regarding manager applications is that they should be listening on UDP port 162 for traps and confirmed inform-requests. In summary, the SNMP distribution model is very simple, in a similar fashion to the whole framework. It serves well enough the centralized management model of LANs and MANs, but it has obvious limitations in more complex environments.

We will now consider a concrete example of SNMP usage in order to demonstrate the access framework. Assume that we would like to find all the routes in the routing table of a network element that "point" to a particular next hop address. The manager application must know the logical name of that network element, which it will map to the network address, typically by using the domain name system. It will then send a SNMP *Get-next* request to port 161 at that address, starting with an "unnamed" route table entry, which will result in the first route entry being returned. This request may be formed as *Get-next(routeDest, nextHopAddr, routeMetric)*. It will then have to repeat this step, passing each time as argument the result of the previous request. Having retrieved the whole table, the manager will have to filter out unnecessary entries and keep those for which the *nextHopAddress* object has the desired value. In SNMPv2, more than one next entries may be requested through the *Get-bulk* primitive, but it might not be possible for the responding agent to fit all of them in the maximum SNMP packet size. In either case, the whole table needs to be retrieved in order to find particular entries, which is expensive in terms of management traffic. In addition, Get-next or Get-bulk requests need as argument the result of the previous request when traversing "unknown" tables. This means the overall latency for performing this operation will be a multiple of the latency incurred for a single retrieval.

Assume also that we would like to be informed if such a new route is added to the routing table, either by management or through a routing protocol. Since traps are used sparsely in SNMP and are also unreliable, the only way to discover new entries is to retrieve periodically the whole routing table. It should be added, though, that after the table has been retrieved once, names of existing entries are known and next traversals can start simultaneously at various "entry" points, reducing the overall latency. Despite that, the management traffic incurred will be roughly the same.

2.3.2. OSI System Management

OSI-SM was designed with generality in mind, and as such it uses a connection-oriented reliable transport. The relevant management service/protocol (CMIS/P) [19] operates over a full seven-layer OSI stack using the reliable OSI transport service. The latter can be provided over a variety of transport and network protocol combinations, including the Internet TCP/IP using the RFC1006 method. The CMIP protocol stack is depicted in Figure 2.6. End-to-end interoperability over networks with different combinations of data link and network layer protocols is supported either through network-level relaying or transport-level bridging as specified in [20]. The upper layer part is always the same and comprises the OSI session and presentation protocols with the Association Control Service Element (ACSE) and the CMIS Element over the Remote Operation Service Element (ROSE) in the application layer [20]. The benefit of transport reliability is outweighed by the fact that a full seven-layer infrastructure is required even at devices such as routers, switches, and multiplexors, which typically run only lower layer protocols. In addition, application level associations need to be established and maintained prior to management operations and the reporting of notifications.

Given the richness and object-oriented aspects of the GDMO object model, CMIS/P can be seen as a "remote method execution" protocol, based on asynchronous message passing rather than synchronous remote procedure calls. The service primitives are a superset of the operations available at the object boundary within agents, with additional features to allow for object discovery and bulk data retrieval, operations on multiple objects, and a remote "retrieval interrupt" facility. The primitives available at the CMIS level are *Get, Set, Action, Create, Delete, Event-report*, and *Cancel-get*. The Get, Set, Action, and Delete operations may be performed on multiple objects by sending one CMIS request which expands within the agent based on *scoping* and *filtering* parameters. Since OSI-managed objects are named according to containment relationships and organized in a management information tree, it is possible to send a CMIS request to a *base* object and select objects contained in that object through scoping. Either objects of a particular level, until a particular level, or the whole subtree may be selected. The selection may be further eliminated through a filter parameter that specifies a predicate based on assertions on attribute values, combined by boolean operators. Scoping and filtering are very powerful and provide an object-oriented database type of functionality in OSI agents. This results in simplifying the logic of manager applications and substantially reducing management traffic.

When applying an operation to multiple objects through scoping and filtering, atomicity may be requested through a *synchronization* parameter. The result/error for each managed object is passed back in a separate packet, which results in a series of *linked replies* and an empty terminator packet. A manager application may interrupt a series of linked replies through the Cancel-get facility. Finally, the Set, Action, Delete, and Event-report operations may also be performed in an unconfirmed fashion. While this is typical for event reports (the underlying transport will guarantee their delivery in most cases), it is not so common for intrusive operations as the manager will not know if they succeeded or failed. Nevertheless, such a facility is provided and might be used when the network is congested or when the manager is not interested in the results/errors of the operation. Figure 2.8 depicts the interactions between applications in manager and agent roles using CMIS (apart from Cancel-get).

The event reporting model in OSI management is very sophisticated, allowing fine control of emitted notifications. Special support objects known as Event Forwarding Discriminators (EFDs) [21] can be created and manipulated in agent applications in order to control the level of event reporting. EFDs contain the identity of the manager(s) who wants to receive notifications prescribed through a filter attribute. The filter may contain assertions on the type of the event, the class and name of the managed object that emitted it, the time it was emitted, and other notification-specific attributes, for example, for an attributeValueChange notification, the attribute that changed, and its new and old values. In addition, an emitted notification may be logged locally by being converted to a specific log record. The latter is contained in a log object created by a manager, which contains a filter attribute to control the level of logging. In summary, OSI management provides powerful mechanisms for dealing with asynchronous notifications and substantially reduces the need for polling. In that respect, it scales much better than SNMP.

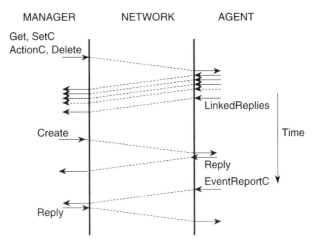

Note 1: Get, Set, Action, Delete may also operate on one only object (single reply).
Note 2: Set, Action, EventReport also have a nonconfirmed mode of operation.

Figure 2.8 CMIS Interactions.

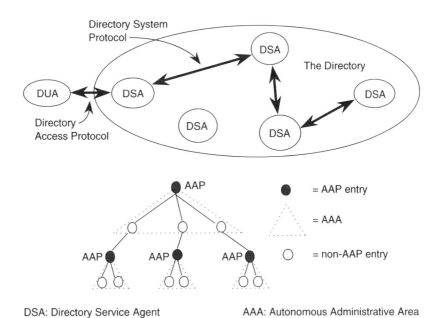

DSA: Directory Service Agent   AAA: Autonomous Administrative Area
DUA: Directory User Agent   AAP: Autonomous Administrative Point

Figure 2.9 X.500 Directory Organizational and Administrative Model.

Distribution aspects in OSI management are supported by the OSI Directory, which provides a federated hierarchical object-oriented database. The Directory resides in many Directory Service Agents (DSAs) that administer parts of the global Directory Information Tree (DIT). Parts of the global MIT belong to different Autonomous Administrative Areas (AAAs) and start at Autonomous Administrative Points (AAPs). DSAs are accessed by applications in Directory User Agent (DUA) roles via the Directory Access Protocol (DAP), while DSAs communicate with each other via the Directory System Protocol (DSP). Accessing the local DSA is enough to search for information anywhere in the global DIT. Figure 2.9 depicts the operational model of the directory and the global DIT. Directory Objects (DOs) are named using distinguished names that express containment relationships, in the same fashion as OSI-managed objects. In fact, the directory naming architecture preceded that of OSI management and was essentially reused in the latter.

OSI management applications, or System Management Application Processes (SMAPs) in OSI parlance, are represented by directory objects. The latter contain System Management Application Entity (SMAE) objects associated with each interface of that SMAP. SMAE DOs contain addressing information as well as information regarding other aspects of that interface, termed Shared Management Knowledge (SMK) [18]. Since the same hierarchical naming architecture is used for both the OSI Directory and management, the two name spaces can be unified. This can be achieved by considering a "logical" link between the topmost MIT object of an agent and the corresponding SMAP directory object.

The universal name space is shown in Figure 2.10 through the extended manager-agent model. The manager application may address objects through

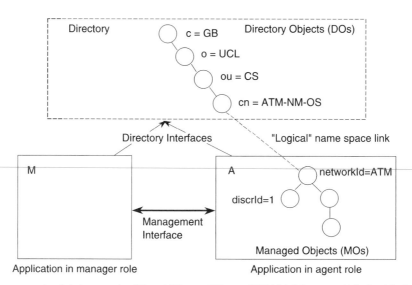

Figure 2.10 The OSI Global Name Space.

their global names, starting from the root of the directory tree, for example, {c = GB, o = UCL, ou = CS, cn = ATM-NM-OS, networkId = ATM, logId = 1, logRecordId = 5}. The underlying infrastructure will identify the directory portion of the name; that is, {c = GB, o = UCL, ou = CS, cn = ATM-NM-OS} will locate the relevant DO and will retrieve attributes of the contained SMAE DO, including the OSI presentation address of the relevant interface. It will then connect to that interface and access the required managed object through its local name, that is, {logId = 1, logRecordId = 5}. Note that the networkId = ATM relative name of the topmost MIT object is not part of the local name. Global names guarantee location transparency, for they remain the same even if the application moves: only the presentation address attribute of the relevant SMAE DO needs to change. Note finally that applications in manager roles are also addressed through directory distinguished names regarding the forwarding of event reports since the destination address in EFDs contains the directory name of the relevant manager.

We will now consider the same example we considered in SNMP in order to see in practice the use of the OSI management access facilities. In this case, the manager application will know the logical name of the device, for example, {c = GB, o = UCL, ou = CS, cn = router-A}, from which the presentation address can be found through the directory. The directory access protocol offers facilities similar to CMIS scoping and filtering. A request will be sent to the SMAP DO with that name, and the psapAddress attribute of the contained SMAE DO will be requested. The manager will then connect to that address and request the relevant table entries through scoping and filtering in the following fashion: *Get(objName = {subsystemId = nw,protEntityId = clnp,tableId = route}, scope = 1stLevel, filter = (nextHopAddr = X), attrIdList = {routeDest, routeMetric})*. The results will be returned in a series of linked replies, sent back-to-back as shown in Figure 2.8. The overall CMIS traffic will be kept fairly low: N linked replies for the matching entries together with the request and the final linked reply terminator packets, that is, $N + 2$ in total. The overall latency will be slightly bigger than that of a single retrieval. It should be added that connection establishment and release are necessary both to the local DSA and the element agent. This does not happen on a per management request basis, but connections may be "cached," as already explained. It should be noted that the discovery of the element agent address through the directory takes place only once.

The manager application would also like to be informed about new route entries "pointing" to the next hop address X. This could be done by using the rich event reporting facilities provided by OSI management. The manager will have to create an EFD with filter *(eventType = objectCreation AND objectClass = routeEntry AND nextHopAddr = X)* and set as destination its own logical name, for example, {c = GB, o = UCL, ou = CS, cn = mgr-Z}. After that, notifications will be discriminated locally within the agent, and the ones matching the filter will be forwarded to the manager. Note that if there is no connection to the manager, the element agent will have to establish it by going through the same procedure and mapping the logical manager name to an address through the directory. The previous observations about connection caching and address mappings are also valid in this case.

2.3.3. OMG CORBA

OMG CORBA was designed as a distributed software infrastructure in which the access protocol is secondary compared to the underlying APIs, or "programming language bindings." Of course, an agreed protocol is necessary in order to achieve interoperability between products of different vendors. The OMG 1.x versions of CORBA specification left completely open the choice of access protocol and concentrated only on concrete programming language bindings. Version 2.0 also specified a Remote Procedure Call (RPC) protocol as the General Inter-Operability Protocol (GIOP) [22]. Two different transport mappings have been defined for the latter, the Internet Inter-Operability Protocol (IIOP) [23] over the Internet TCP/IP (shown in Figure 2.6) and the DCE Common Inter-Operability Protocol (D-CIOP). The ODP *access* transparency prescribes independence of the underlying access protocol, and CORBA provides both independence and portability due to the agreed APIs. The access protocol could change without any effect on application-level software!

The agreed CORBA protocol is a connection-oriented reliable RPC that uses TCP and IP as transport and network protocols, respectively. Applications that use CORBA-based communications are guaranteed transport reliability, in a similar fashion to OSI management and unlike SNMP. The CORBA RPC protocol is a *request/response* type of protocol in which the exact structure of the request and response packets is defined by the IDL specification of the accessed CORBA interface. No special facilities are built in the protocol for object discovery and multiple object access in a similar fashion to the SNMP get-next, get-bulk or the OSI management scoping and filtering. Instead, such facilities are provided in a limited fashion by the ORB and by special *servers*. In summary, the CORBA RPC protocol provides a *single* object access mechanism with higher-level facilities provided by standard OMG servers [16].

The CORBA operational paradigm is different from that of OSI and Internet management, for it originates from the distributed system world. CORBA objects are specified and accessed separately, in contrast to the managed object cluster administered by an agent. Another key difference is that CORBA objects are most commonly addressed by *type* and not by *name*. This is due to the nature of distributed systems where, typically, instances of the same type offer exactly the same service, for example, printer servers, statistical calculation servers, and so on. Of course, this does not mean that there are no support mechanisms to distinguish between instances of the same type (name servers, traders). It means, however, that the whole framework is optimized toward a *"single object access, address by type"* style of operation, in contrast to the manager-agent model which is optimized for *"multiple object access, address by name"* style of operation.

A CORBA object instance can be addressed by type through the ORB, in a fully location-transparent manner. The ORB will find an instance of that type and return an object reference to the client object. If there are many instances of that type in the ORB domain, many references will be returned. Instances of the same type can be distinguished through naming servers or traders. A naming server [16] can be used to map a name to an interface reference. When an object instance is created, the naming server needs to be "told" of the mapping between the object's name and its interface reference. Subsequently, client objects can resolve object names to object references through the naming server. We should recall here that the OMG naming architecture is very similar to that of OSI management/directory, but objects may have more than one names.

A trader [24] supports more sophisticated queries, matching sought properties of the target object(s). Objects can export their interfaces to the trader together with a list of *attributes* and a list of *properties*. Clients may request the object references of a particular type that match assertions on attributes and properties. The difference between the latter is that attributes may change dynamically, while properties are fixed during the lifetime of an object instance. As such, the trader needs to evaluate assertions on attributes by retrieving them from all the instances of the type associated with the query. The function of the trader is very similar to filtering in OSI management. A key difference is that only interfaces of a particular type can be searched through the trader. An additional difference is that filtering is tightly coupled with OSI-managed objects through the supporting agent, while the ODP/OMG trader is a separate server. Finally, traders can be in principle federated in order to cope with big object spaces and different administrative domains.

Notifications in ODP/OMG are supported by *event* servers. Emitting and recipient objects need to register with the event server, and special objects called *channels* are created and managed for every type of notification. Emitting objects invoke an operation on the relevant event channel, while the notification is passed to registered recipient objects either by invoking operations on them (*push* model) or through an operation invoked by the recipient object (*pull* model). There is no filtering as in OSI EFDs, while event servers can be in principle federated for scalability and interdomain operation. The key difference with OSI management is the lack of fine grain filtering, which results in less power and expressiveness and more management traffic. OMG is currently working toward the specification of *notification* servers which will provide filtering and will also take over the management of channels, providing a higher-level way to deal with notifications.

We will now examine how CORBA could be used for network and service management, contrasting its approach to the protocol-based OSI and Internet management approaches. But let's first recapitulate the operational paradigm of the latter. Managed elements or management applications that assume an agent role provide management interfaces. A management interface consists of the formal specification of management information and of an access service/protocol that is mapped onto a well-defined protocol stack. While the management information specification provides the MIB *schema*, object discovery and multiple object access facilities allow applications in manager roles to discover dynamically existing object instances. Operations to objects are always addressed through the supporting agent, which provides query facilities in a database-like fashion. In addition, the agent discriminates emitted notifications according to criteria preset by managers. Applications may discover each other through the directory in OSI management, while predefined addresses are used in SNMP.

If CORBA is used as the underlying access and distribution mechanism, managed objects can be mapped onto CORBA objects, accessed by client objects in managing roles. The key difference is that clusters of managed objects logically bound together, for example, objects representing various aspects of a managed network element, are not seen collectively through an agent. As such, an important issue is to provide object discovery and selection facilities similar to OSI scoping and filtering. Such facilities are very important in management environments where many instances of the same object type typically exist, with names not known in advance, for example, call objects. Facilities similar to scoping are not currently supported in CORBA, but it should be possible to extend name servers to provide similar functionality since they maintain the logical name space. Facilities similar to OSI filtering, as explained above, are currently provided by traders, but are not as powerful. An alternative solution would be to provide special *query* servers, offering object selection facilities based on scoping and filtering, in a similar fashion to OSI management.

The problem with the use of CORBA as described above is that federation is a key aspect in order to achieve scaleable systems. In essence, it will be necessary to have dedicated name servers, traders, and event/notification servers for every logical cluster of managed objects, for example, in every managed element, in order to reduce traffic and increase real-time response. Those "low-level" servers will be unified by "higher-level" servers in a hierarchical fashion, but federation issues have not yet been worked out and are not simple. In addition, even with such facilities in place, the management traffic in terms of the required application-level packets will be at least double compared to that of OSI management. In CORBA, matching object references will be returned to the client object, and the operations will be performed on an object-by-object basis. In OSI management, the multiple object access request will be sent in one packet while the results will be returned in linked replies, one for each object accessed. The use of CORBA for network management by using federated trading is depicted in Figure 2.11.

We will now consider the same example we considered in SNMP and OSI

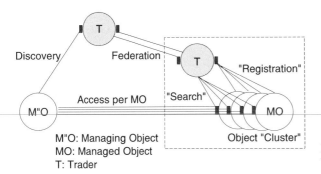

M"O: Managing Object
MO: Managed Object
T: Trader

Figure 2.11 The Use of CORBA for Network Management.

management in order to see in practice the CORBA access facilities. The manager in this case will be a CORBA client, which will have to discover the right routing table entries through the trader. If there is no federation, a central trader will be used, and objects such as route table entries will need to export their properties to it across the network. One of their properties will have to be the logical name of the router so that assertions about routes in different router nodes are possible. If there is federated trading, a trader could be located at the router node so that exporting properties by local objects would not generate any traffic. The manager will contact a trader and perform an operation to import interfaces with particular properties. The properties, in a filter-like notation, would be *(ifType = routeEntry AND router = router-A AND nextHopAddr = X)*. This will result in a number of routeEntry interface references returned to the manager. The latter will then have to perform a Get operation for each entry and retrieve the *routeDest* and *routeMetric* attributes. These operations may be performed concurrently, so that the overall latency will be similar to that of one operation. Note, however, that the synchronous nature of RPC necessitates the use of a multi-threaded execution environment, with a separate thread for every invocation. The management traffic incurred will be 4 RPC packets for trading (2 to/from the domain trader and another 2 between the latter and the trader in the router); and 2*N RPC packets for retrieving the entry attributes, that is, $2*(N + 2)$ packets in total.

Event operation in CORBA is less powerful than in OSI management but nevertheless useful for avoiding polling-based management. Assuming a *routeEntryCreation* event is defined, the manager will have to register with the event server in order to receive this event. Typically, every event will involve 4 RPC packets: 2 between the emitting object and the event server and 2 between the latter and the manager. Since OMG event services do not support filtering, the manager will receive events for all the new route entries in all the routers and will select locally those of interest. Federated notification servers will be necessary in the future to provide more sophisticated event management facilities.

2.3.4. Summary and Comparison

In summary, SNMP adopts a connectionless unreliable transport while both OSI management and CORBA adopt a connection-oriented reliable transport paradigm. The only difference between the latter two is that connection establishment is "hidden" in the case of CORBA through the ORB, while it may be presented to applications in the case of OSI management. The main reason for the connectionless (CL) approach in SNMP is simplicity in managed elements, with complexity shifted to manager applications that have to achieve reliability through retransmission. Experience has shown that it is very difficult to optimize retransmission in the same fashion this is done by reliable transport protocols such as the Internet TCP and the ISO/ITU-T TP. In addition, compact implementations of reliable transport stacks have become a commodity, while the recent advances in inexpensive memory and processing capabilities suggest that the argument of simplicity is no longer valid. Finally, the emerging broadband technologies (SDH/SONET transmission, ATM switching) are connection-oriented, which means they are better aligned with the OSI-SM and CORBA approaches.

Coming to the access paradigm, SNMP and OSI management adopt the manager-agent approach with managed object clusters visible across a management interface and query/event/multiple object access facilities, while CORBA relies on a single-object access paradigm, with special servers providing additional facilities. The query, event, and multiple-object access/bulk data transfer facilities of OSI management are very powerful, for they have been designed specifically for telecommunications network management and have not been compromised in order to reduce agent complexity. The same facilities in SNMP are less powerful, and they result in less expressive power and much more management traffic. On the other hand, they keep element agents simple and have resulted in making SNMPv1 a success, at least for private (LAN/MAN) networks. In OSI management and SNMP, there are no implementation constraints in the sense that they are both communications frameworks. As such, compact optimized implementations are possible.

OMG CORBA, on the other hand, projects an object-oriented distributed software framework that is not specific to management and, as such, more general. Facilities similar to those provided by OSI/SNMP agents may be supported by OMG servers: name resolution and object selection will be supported by name servers, sophisticated filtering by traders, while events are currently supported by event and, in the future, by more sophisticated notification servers. Given the fact that in management networks there will exist hundreds of thousands managed objects, federation is absolutely important for scalability and timely responses. The use of facilities such as name, event/notification servers, and traders for network management is currently a research area, while federation issues have not yet been resolved. An additional issue is the complexity of the overall resulting framework as CORBA dictates conformance to internal software interfaces, which leaves less space for optimized implementations. For example, the feasibility of network elements with tens of thousands of CORBA managed objects needs to be investigated.

2.4. VARIOUS OTHER ISSUES

2.4.1. Scalability, Flexibility

Scalability is an issue when managing large networks. Management services are used by the network operator and are transparent to end-users. Management traffic should be kept low so that most of the network bandwidth is available to end-user services. Obviously, choice of management paradigm (event-driven, polling-based) has an impact on the amount of management traffic. We will examine a simple case study in order to quantify management traffic and assess scalability issues.

SNMP is particularly well-suited for LAN and MAN environments that are inherently connectionless, the available bandwidth is relatively high (10 to 100 Mbits/s), and the error rate is very low. In such networks, the overhead of polling is only a small fraction of the available bandwidth. No sophisticated retransmission is necessary because the small error rate allows communications to take place essentially over one "link." We will assume a modest LAN/MAN cluster consisting of 100 routers with an average of three interfaces per router and a maximum latency of five minutes for detecting "interface down" alarm conditions. This requires (100*3)/ (5*60) = 1 polls/sec to monitor the up/down status of interfaces, or 1/3 polls/sec if all the requests for a router are combined in a single SNMP packet. Adding to this 5000 workstations/PCs/servers whose status needs to be known at an average latency of 10 minutes, we need another 5000/(5*60) = 8.33 polls/sec, the grand total being about 10 polls/sec. Since an SNMP packet is about 500 bytes, this results in a management bandwidth of 0.1 Mbits/s, or 1 percent of the total Ethernet bandwidth (10 Mbits/s), which is affordable.

The above calculations concern polling to determine only the rudimentary status of the network, in terms of its most important components. Adding to this system and application management, for example, operating system load/ users, terminal servers, database management systems, the domain name system or directories, mail systems and so on, including performance and accounting issues in addition to fault detection, and the above figure will be much bigger. It is obvious that doing the same thing over a wide area network with a lot of "thin" point-to-point links and higher probability of congestion will result in a lot of additional load, deteriorating the network's overall health. In such an environment, sophisticated retransmission mechanisms will also be necessary because the probability of packets being lost will be much higher. The solution for scalability is event-driven management, with facilities such as event management with filtering, event logging, metric monitoring with thresholding, and summarization (see section 2.4.2). SNMP does not provide such facilities, at least for element management, while OSI management does. OMG CORBA, on the other hand, was not designed specifically for management; its event model is not as powerful, and it lacks generic management facilities.

We will now examine suitability for hierarchical management. By hierarchical management, we mean a management system organization in which management applications are organized in a logical layered fashion, with applications in higher layers being shielded from unnecessary detail and having a global view of the network, services, or policies. A hierarchical management structure was depicted in Figure 2.2 and is exemplified by the TMN model. The SNMP framework was designed to allow management capabilities to be fielded in the largest possible number of network elements. As such, simplicity dictated the use of connectionless transport, no sophisticated event facilities, and a rather crude information model. Such design decisions address mainly the lowest level of a management hierarchy (element management). However, when it comes to hierarchical management, issues such as management application size, complexity and processing requirements become largely irrelevant. Management applications in this layered hierarchy usually operate in powerful workstations. In this case, the simplicity of the SNMP framework becomes a liability as it restricts the available expressive power and introduces limitations.

Finally, we will consider flexibility and suitability for distributed application and service management. Both SNMP and OSI management may be used for distributed application management. On the other hand, distributed applications should be managed through dedicated agents, and both SNMP and OSI agents are too complex to be "bundled" together with them. In general, it is natural to manage distributed applications employing the same technology used to build them in order to achieve reusability and economies of scale. Internet applications can be managed with SNMP, OSI applications can be managed with OSI, and CORBA applications are best managed through CORBA. Given the fact that CORBA was conceived as a mechanism to build distributed systems, it is best to build and manage new distributed applications through CORBA.

The same is true for service management regarding new advanced services, for example, video-conferencing and joint document editing. In this case, it is difficult to differentiate between service operation and service management. For example, subscription management to a new advanced service is a management activity that is closely related to the operation of that service. This is exactly the thinking behind the adoption of CORBA as the basis for the TINA DPE, as the TINA architecture tries to unify service operation and service management mechanisms and procedures. In this unified model, a video-conferencing bridge can be seen as a CORBA object with both service and management interfaces as opposed, say, to an object with a service interface and an associated OSI agent for TMN-based service management. The two approaches are depicted in Figure 2.12. In summary, OMG CORBA is a more flexible mechanism than SNMP and OSI management for *managing* distributed applications because it is also a mechanism for *building* them in the first place.

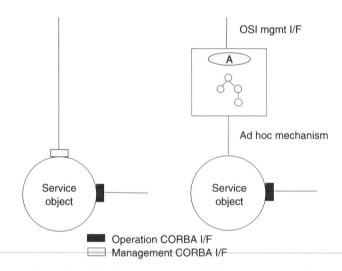

Figure 2.12 Service Operation and Management Models.

2.4.2. Generic Management Functionality

One fundamental difference between the Internet and OSI management frameworks is that the Internet follows a "lowest common denominator" approach, resulting in very few common object specifications that should be globally supported. This approach is in line with its fundamental axiom which dictates simplicity in managed elements. On the other hand, a number of generic management functions are standardized in OSI management in order to provide a well-defined framework for dealing with common tasks and achieving reusability. These specifications emanate from the five functional areas (Fault, Configuration, Accounting, Performance, Security—FCAPS) and are collectively known as the System Management Functions (SMFs) [25]. The notion of generic functionality in CORBA is supported by the Common Object Services [16]. Since CORBA was not designed specifically for management, it only partly supports functionality similar to the OSI SMFs. In this section, we examine generic management functionality through the OSI SMFs and compare it to similar facilities in the SNMP and CORBA frameworks.

There are three types of OSI SMFs:

1. Those that provide *generic* definitions of object classes, or simply attributes, actions, and notifications for common tasks.
2. Those that provide *system* definitions which complement the management access service by providing a controlled mechanism to deal with notifications (Event Reporting/Dissemination, Log Control).
3. Those that provide *miscellaneous* definitions; we could currently group here the security-related functions (Access Control Objects, Security Alarm Reporting, and Security Audit Trail).

Starting first from the third category, such functions exist in the SNMPv2 security framework (apart from the security audit trail). The SNMPv2 Party MIB has similar functionality to the OSI Access Control Objects. These facilities are of paramount importance for the security of management and are absent in SNMPv1. The OMG CORBA security framework supports access control and security audit functions (see section 2.4.3).

The second category is extremely important, for it provides the means for event-driven instead of polling-based management. The philosophy of both SNMPv1 and v2 is based on polling, at least between element managers and NEs. As such similar facilities exist only partially, as already discussed. The event group of the "manager-to-manager" MIB provides facilities similar to OSI event reporting but without filtering, while the same is true of the OMG event service. Logging services are not provided, in either the SNMP or OMG CORBA frameworks.

The first category provides a host of functions that support generic functionality. The first and most important of those are Object Management, State Management, and Alarm Reporting. Object Management provides three generic notifications related to configuration management that all OSI-managed objects should support: *object creation*, *object deletion*, and *attribute value change*. State Management provides a number of generic *state* attributes (administrative, operational, usage state, etc.) and a *state change* notification. It also prescribes state transition tables according to the state model. Finally, alarm reporting provides a set of generic *alarm* notifications: quality of service, communications, equipment, environmental, and processing error alarm.

Other MIB specifications should use the above definitions in order to model object, state, and alarm aspects. Generic configuration, state, or alarm managers can be written in a fashion that makes them independent from the semantics of a particular MIB. For example, a configuration monitor could be an application that connects to managed elements and requests all the object creation, deletion, attribute value, and state change notifications in order to display changes to the human manager. Such an application can be written once and reused as it only needs to be "fed" the formal specification of the element MIBs in order to be able to display meaningful names for the objects emitting those notifications. OSI management platforms typically provide a set of generic applications that are based on those common specifications. SNMP and CORBA do not provide similar generic facilities, but CORBA may reuse the OSI ones if the relevant GDMO specifications are translated to CORBA IDL, as described in the next section. This observation also holds for the rest of the OSI SMFs, which are described next.

Monitor Metric objects allow the observation of counter and gauge attributes of other MOs and their potential conversion to *derived gauges*, which may be statistically smoothed. The latter have associated threshold and tidemark attributes that fully support event-driven performance management capabilities, relegating "polling" within a managed element. The SNMPv2 manager-to-manager MIB offers a similar facility but without statistical smoothing or the possibility of combining different attributes in order to produce a comparison rate, for example, for error versus correct packets. Summarization objects allow a manager to request a number of attributes from different objects of a remote system to be reported periodically, possibly after some statistical smoothing. These attributes can be specified using scoping and filtering while intermediate observations may be "buffered." This facility is important for gathering performance data for capacity planning and is typically used together with logging. SNMP does not provide such a facility.

Accounting Metering provides generic objects to support data collection for resource utilization. Test Management defines generic test objects to provide both synchronous and asynchronous test facilities, modeling generic aspects of testing and separating them from specific test aspects. Scheduling Management provides generic scheduler objects that could schedule activities of other MOs which support such scheduling on a daily, weekly, monthly, or other periodic basis, for example, event forwarding discriminators, and logs. Response Time Monitoring supports performance management by allowing the measurement of protocol processing time and network latency between systems. Time Management permits the delivery of correct time and synchronization of the clocks of distributed systems. Software Management permits the delivery, installation, (de-)activation, removal, and archiving of software in a distributed fashion. SNMP does not provide similar generic facilities.

In summary, SMFs provide useful generic facilities that most systems should support, enforcing a common style of operation that can result in generic managing applications or simply generic managing functions. SNMPv2 matches partly the event reporting and metric objects, and these only in the manager-to-manager domain. OMG CORBA provides less powerful event reporting facilities, while the rest of the OSI SMFs could be translated to CORBA IDL and used in CORBA environments. Despite this theoretical possibility, this approach has not yet been put into practice.

2.4.3. Security

Security of management is of paramount importance, especially for *intrusive* operations that result in the modification of management information. Security is particularly important across different administrative domains but is also necessary within a domain, especially in cases where that domain is open to external management traffic. Despite its importance, it has taken a long time to produce agreed-upon workable solutions. A common aspect in all these frameworks is that security mechanisms have been almost an after-thought, after the main aspects have been standardized and nonsecure implementations have existed in the marketplace for some time.

First we will describe security threats and then security services used to protect against those threats. A third-party application may attempt to subvert the management interaction between a pair of communicating management applications by:

- Masquerading as a legitimate application and then performing unauthorized management operations.
- Modifying information while in transit.
- Reordering or replaying messages in transit.
- Capturing (and examining) confidential management information in transit.

The security services used to protect against these threats are the following:

- Peer entity authentication, which establishes unambiguously the identity of the initiator of an operation and is an essential input to an access control decision function.
- Data origin authentication, which provides an assurance guarantee that data really do come from where they seem to.
- Connectionless integrity, which ensures that management PDUs cannot be modified without detection.
- Stream integrity, which guards against misordering PDUs in a stream (including replays).
- Confidentiality, which prevents capture and examination of management information.
- Access control, which enables one to discriminate between different managers or client objects regarding the operations they are allowed on managed objects.

We will now consider the approaches taken for providing these services in the three management technologies. In the Internet management world, SNMPv1 has little security in the form of "password-based" authentication and access control [2].

There is a special field in SNMPv1 packets that identifies a *community* name. Every agent needs to know in advance the names of various communities with different access rights. As such, MIB access is restricted based on that name. The default community is called *public* and provides a minimal level of access, for example, *read-only* for objects that can be accessed by anybody. The first problem with this scheme is that it does not allow for the dynamic configuration of agents with respect to communities and access rights. The second and most important problem is that the community name is passed across unencrypted, which makes the scheme vulnerable to *capturing* attacks. Because of its simple-minded nature, the SNMPv1 security mechanism is not trusted and, effectively, not used. In the absence of strong security mechanisms, many SNMPv1 manageable devices do not implement Set operations in order to avoid the potentially disastrous effects of malicious attacks. Even worse, the various IETF groups involved in the definition of new MIB specifications have refrained from allowing extensive intrusive management capabilities through Set operations. The absence of security has resulted in the use of SNMPv1 as a remote *monitoring* rather than a *management* framework.

The initial Internet SNMPv2 RFCs, published in 1993, included a security framework, but the relevant IETF working group never reached agreement regarding the security aspects. As such, the new version of the SNMPv2 RFCs, published in 1996, has made the security framework optional and allows for the possibility of multiple security frameworks. The simple community-based scheme can also be used with SNMPv2. A more comprehensive security framework caters for the attacks mentioned above by providing comprehensive security services [26].

In SNMPv2, the relevant communicating entities are referred to as *parties*. Peer-entity, data-origin authentication, and connectionless integrity are supported by a scheme based on the Message Digest 5 (MD5) algorithm. With MD5 authentication, each party is associated with a secret key, held securely at source and destination. The source party applies the MD5 algorithm to a combination of PDU and secret key in order to generate a security checksum that is appended to the PDU. The destination repeats the calculation to verify the source party's identity. The use of shared secret keys introduces a problem of key distribution. This can be addressed by secure key exchanges through SNMPv2 itself. However, this can only be done if it is certain the security of the key exchange channel itself has not been compromised. In addition, there is a bootstrap problem to solve; that is, keys will need to be distributed *out-of-band* for the first time. Stream integrity may be supported by the time-stamps included in every message, assuming clocks are synchronized. An alternative novel approach has been proposed for SNMPv2 which uses time as perceived by the agent only, and, as such, it avoids clock synchronization. Confidentiality is supported by encrypting portions of an SNMPv2 message using the Data Encryption Standard (DES). Finally, access control is provided through the *party* MIB which describes the access rights of different parties to objects in the agent's MIB.

In OSI, security services other than access control are applicable to all the application service elements and not just to CMISE. For example, the same authentication, integrity, and confidentiality services can be used for management, directory access, and file transfer. Authentication services were initially developed for the OSI Directory, but the need for a general security framework led to the Generic Upper

Layer Security (GULS) ITU-T recommendations [27]. These were completed in 1996, while specific lightweight profiles have also been produced by workshops to ease their acceptance and introduction into the marketplace. A key difference between the OSI and SNMP security frameworks is that the OSI also defines mechanisms for asymmetric public-key cryptography in addition to symmetric secret-key-based schemes. In asymmetric schemes, the source application needs to demonstrate knowledge of its own *secret* key. This requires a certain amount of infrastructure support in the form of *certification authorities* which vouch for the bindings between the directory name of that application and its *public* key. There is no key-distribution problem as each entity holds (securely) its own secret key while it "advertises" its public key to the directory. Asymmetric public key cryptography requires the RSA algorithm, which can be computationally expensive when performed in software while implementations in hardware (*smartcards*) overcome this limitation. Directory access by a management application is necessary for both advertising its own public key and obtaining a peer entity's public key.

OSI GULS services include authentication, connectionless/stream integrity, and confidentiality. The default digest and encryption algorithms used are MD5 and DES, respectively, but any other algorithm can be used after negotiation. Two lightweight profiles have been defined by the ANSI T1M1 and the IEEE Open Implementers Workshop (OIW), the *mini-* and *micro-GULS*, respectively. These may operate using secret-key authentication, while key distribution aspects are deliberately left unspecified. The key difference between the two profiles is that micro-GULS supports the encryption of a whole PDU only, while mini-GULS supports the encryption of selected PDU fields, providing additional flexibility and increased efficiency. While GULS requires a presentation layer PDU transformation and, as such, protects all application layer services, an alternative approach is to provide security services based on ROSE PDU transformation [29]. Such an approach is simpler since it does not require the modification of the OSI presentation layer. On the other hand, it can only protect ROSE-based application services. The ANSI T1M1 and IEEE OIW have also produced ROSE-based security service specifications as a simpler alternative to GULS. While using GULS or ROSE-based security services, access control is provided through special managed objects that protect other *target* objects or even individual attributes, actions, and notifications in a very flexible manner. Management of security facilities is provided through security alarm reporting and security audit trail functions [25].

OMG CORBA security services are a fairly recent addition to the overall framework. The OMG security framework is very broad and supports a plethora of possible security services and mechanisms. The current security specification defines APIs that provide access to security services supported by a number of different, potentially replaceable security architectures and policies. The security services accessible through those APIs include authentication, protected message exchange, that is, integrity and confidentiality, access control, security auditing, and delegation of security rights to intermediate authorities [28]. In fact, the CORBA security approach offers a "shopping-list-oriented" solution space to which implementations of underlying security architectures and policies can adhere. Accordingly, interoperability is not the focus of this high-level specification. In order to foster interoperability, however, a set of profiles specifying particular security services and relevant supporting mechanisms have been defined. A simple negotiation protocol has been defined in order to choose a particular profile between ORBs.

In summary, OSI management and OMG CORBA security solutions have been fully specified and secure implementations of relevant products are expected to appear in the marketplace soon. On the other hand, SNMPv2 has opted for an optional security framework whose adoption is questioned.

2.5. INTERWORKING AND COEXISTENCE

In this section, we examine interworking and coexistence aspects for the three different technologies. We look first at interworking and coexistence between OSI and Internet management in the TMN context, where SNMP-capable network elements may need to by managed by TMN applications. We then look at interworking between OMG CORBA and SNMP in the TINA context, where network elements need to be managed by CORBA objects. Both of these cases are unidirectional in the sense that SNMP is only considered in the managed end of the spectrum. Finally, we examine interworking and coexistence between the OSI management and OMG CORBA in both directions: first in the OSI management to CORBA direction, which is necessary to manage CORBA-based distributed applications from a TMN environment; and then, in the CORBA to OSI management direction, which is needed to manage network elements with Q interfaces or to reuse existing TMN management services in a TINA environment.

2.5.1. OSI and Internet Management

Interworking between OSI management and SNMP is mostly necessary to manage SNMP-capable network elements in a TMN fashion. This is particularly common for ATM equipment, as relevant SNMP information models have been available for some time before the relevant TMN recommendation, resulting in a number of SNMP-capable ATM elements in the marketplace. In transmission technologies such as SDH, the situation has been the reverse with early Q-compliant available elements. In general, however, it is expected that in the short to medium term (i.e., for the next few years), it will be necessary to manage SNMP-capable elements from a TMN environment.

Interworking between CMIS/P and SNMP has led to a lot of research trying to bridge the two worlds. Various solutions have been proposed, all of which can be classified into two broad categories:

1. Integration in the manager end
2. Integration in the agent end of the manager-agent model

The integration in the manager end means that one accepts the diversity in the supported technology by managed elements and tries to provide element management applications that understand the different underlying information models and access mechanisms. This approach is often referred to as *dual stack manager*, since these applications will need to understand both the OSI and Internet management models. Such an approach has been envisaged by the X/Open Consortium in providing the XOM/XMP [30] Application Programming Interface (API), which provides uniform access to both CMIS and SNMP services.

Such an approach is suspect, however, because it is difficult to conceal which model the managing application deals with since both the underlying information models and access mechanisms have important differences. For example, the nature of objects and the naming schemes are different in the two models. This is also the case with respect to the supported communication and access paradigms. Of course, this does not mean that such an approach is not feasible but simply that integration cannot be seamless, increasing significantly the development effort and investment for dual manager applications. The dual stack manager approach is depicted in the left part of Figure 2.13.

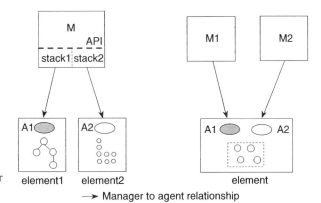

Figure 2.13 The Dual-stack Manager and Dual-stack Agent Approaches.

Integration by agents can again be classified into two broad categories:

1. The dual agent approach
2. The application gateway approach

By *dual agent* we mean that two agents should exist for every managed element, both an SNMP and an OSI one. The information models will be semantically similar, which implies that the associated "real resource" aspects could be the same. As such, investment in providing such agents could be reduced if a modular approach were followed. In the latter, managed objects are realized in a "model/protocol independent" fashion and are associated with both SNMP and CMIS/P access methods.

This approach is depicted in the right part of Figure 2.13, in which the objects in the dual agent are model independent, with different views presented through the two agents. Despite the fact that this approach is technically feasible, it requires heavy investment and additional resources in managed elements. As such, no products support this type of functionality to date.

The *application gateway* approach provides the most promising and powerful solution for integrating the two frameworks. In this, an application acts as a gateway (*proxy* or *adapter* are two other terms often used) for one or more agents of the other framework, exporting "converted" information models and providing service conversion from one access method to the other. The conversion between the information models can be performed either manually or automatically. *Manual conversion* means that human heuristics may be applied to result in an "elegant" model. In many cases, the target model for that technology may already exist, in which case a gateway should simply map one to the other. *Automatic conversion* means that a well-defined set of rules exists and can be used to automate translation of any MIB specification from one model to the other. As a result of automatic conversion rules, the dynamic interaction translation and subsequently application gateways may be automated.

Automatic conversion is usually unidirectional: it can only be bidirectional if the two information frameworks are equally powerful and expressive. This is not the case with OSI management and SNMP, GDMO being much more powerful than the SNMPv1/v2 SMI. As such, there may be automatic conversion of an SNMP information model to the equivalent GDMO one but not vice versa. Human intervention is required for mappings in the opposite direction. For example, there is no deterministic method for emulating a CMIS/P *Action* through an SNMP *Set*.

Research effort led by the Network Management Forum (NMF), known as the ISO/ITU-T and Internet Management Coexistence (IIMC) work, resulted in a set of NMF documents that provide rules for automating the SNMP to GDMO information model conversion and for building generic application gateways between CMIS/P and SNMP [31]. The automatic conversion between the two frameworks relies on the simple observation that the SNMP structure of management information is a pure subset of the OSI one. SNMP objects are equivalent to OSI attributes, groups are mapped to classes, and table entries become separate classes. Traps are mapped to notifications associated with a *cmipSnmpProxyAgent* class that represents the proxied SNMP element. Based on those rules, one can fully automate the CMIS/P to SNMP service conversion. Commercial products providing this functionality already exist. Typically, every time a new element with an "unknown" MIB needs to be adapted for, an off-line procedure is involved to let the gateway "know" of this MIB through suitable translators/compilers. The latter generate run-time support in a data-driven fashion so that the gateway logic does not need to be altered. The generic gateway is a Q-Adapter in TMN terms. The application gateway approach is shown in Figure 2.14.

The important aspect of the generic gateway approach is that investment is rather small compared to the end result, which is OSI manageability of any SNMPv1/v2 capable element. The key benefit OSI management brings to the SNMP world is event-driven management through the Systems Management

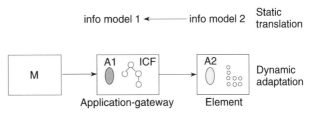

Figure 2.14 The Application-Gateway
Approach.

ICF: Information Conversion Function
→ Manager to agent relationship

Functions. For example, metric and summarization functions, together with event reporting and logging, may be used to provide sophisticated management capabilities, eliminating polling in the local environment between the gateway and the proxied SNMP agent. The benefits of the generic gateway approach are described in [33]. The key drawback, on the other hand, is that the resulting information model does not exploit the object-oriented aspects of GDMO: inheritance is only two-level (every class inherits only from *top*) while containment is also fairly "flat." Furthermore, the resulting information model needs to be standardized in order to be considered a standard Q interface in TMN terms.

2.5.2 ODP/OMG CORBA and Internet Management

Interworking between OMG CORBA and SNMP is necessary in order to manage SNMP-capable network elements in a TINA environment. Early TINA prototypes have been using nongeneric adapters supporting only the necessary functionality (e.g., for Connection Management). As other parts of the TINA management architecture, such as Resource Configuration Management, are further specified and expanded, the need to access SNMP-capable elements will be much greater. If in the long term TINA plans to provide full-scale TMN-like functionality, SNMP or Q-capable network elements will need to be accessed by CORBA managing objects. A generic approach to information model translation and dynamic adaptation will pay dividends, for it will minimize the necessary investment and will allow reuse of the relevant adapters. The need to interwork between OMG CORBA and both Internet and OSI management led to the joint effort between X/Open and the NMF, known as the X/Open Joint Inter-Domain Management (XoJIDM) task force [34]. In this section, we concentrate on the issues behind OMG CORBA and SNMP interworking.

CORBA IDL is a more powerful object-oriented interface specification language than the SNMPv1/v2 SMI templates. In addition, the NMF IIMC work for mapping a SNMP SMI model to the equivalent GDMO one is of direct relevance, and the same modeling principles apply for translation to CORBA IDL. SNMP objects can be mapped onto IDL interface attributes, groups can be mapped onto IDL interfaces, and table entries can be mapped onto separate IDL interfaces. Finally, traps become notifications modeled by two IDL interfaces: a *Notification* interface that should be inherited by any managing object wishing to receive notifications according to the push event model; and a *NotificationPull* interface that should be inherited by notification server objects supporting the pull event model.

Every translated IDL interface inherits from a *SmiEntry* base interface, which in turn inherits from CORBA's *Object*, as do all IDL interfaces (see Figure 2.15). *SmiEntry* provides generic SNMP-related functionality in the form of a "naming" attribute and other generic aspects. While SNMP is not particularly powerful as an access method when compared to CMIS/P, it still offers some access facilities that cannot be easily provided by CORBA. For example, one SNMP request may retrieve or change the value of attributes across different table entry instances, for example, the status of interfaces at a particular node and the next hop address of routes. In CORBA, objects are seen as different entities through IDL interfaces, and, as such, a separate method invocation is needed for each interface.

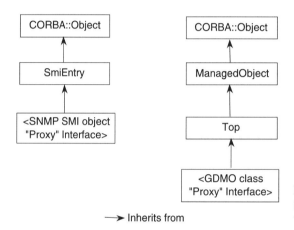

Figure 2.15 Inheritance Hierarchy from SNMP SMI and GDMO to IDL Translation.

2.5.3. ODP/OMG CORBA and OSI Management

Interworking and coexistence between OSI management and OMG CORBA is needed in both TMN and TINA environments: it should be possible to manage CORBA-based distributed applications from a TMN environment (e.g., in the context of service management), and it should be possible to access TMN-compliant elements from a TINA environment or to reuse existing TMN-based management services. Some observers see this latter case as a possibility for a TMN to TINA coexistence strategy: the TMN network layer management services could be reused, with the TMN service layer being replaced by equivalent TINA functionality.

In order to manage CORBA server objects through OSI management, we need first to translate IDL to GDMO/ASN.1 and then to provide mappings between the CMIS/P and the CORBA access mechanisms. Mapping CORBA IDL interface definitions to GDMO classes is fairly straightforward since IDL is simpler than GDMO. IDL attributes are mapped onto GDMO attributes, IDL methods are mapped to GDMO actions, and IDL interfaces to GDMO classes. CORBA object references and names will be mapped onto OSI distinguished names. The generic application gateway needs to interact with standard OMG services in the CORBA domain, for example, the OMG Name Service to resolve distinguished names to object references, the OMG Lifecycle Service to create new object instances, and the OMG Event Service in order to receive events and forward them to interested OSI managing applications. Scoping and filtering can be resolved within the gateway, with one CMIS request mapped onto one or more requests on IDL interfaces. This type of gateway can also be conceived as an OSI agent for which the real resources associated with the managed objects it administers happen to be CORBA objects.

Mapping in the opposite direction is a more difficult proposition. GDMO/ASN.1 as an information specification language and CMIS/P as the access method have a number of aspects for which there exist no IDL and CORBA equivalents. These include the late binding of functionality to managed object instances through the use of conditional packages; the existence of notifications as part of managed object specifications; the fine grain support for event discrimination; and the use of scoping and filtering as "query language" facilities that may result in multiple replies. In addition, a GDMO action on a single managed object instance may also result in multiple replies (e.g., a testing action taking a long time to execute with periodic results). It should also be noted that GDMO attributes cannot be mapped directly onto IDL attributes since user exceptions with specific error information may be raised as a result of access to them. In IDL it is not possible to associate user exceptions with attribute access.

Despite these differences, it is still possible to use workarounds in order to achieve a generic mapping. GDMO attributes may be mapped onto access methods specific to the attribute in hand, according to its property information (e.g., *administrativeState_get*, *administrativeState_set*). GDMO actions resulting in single replies may be naturally mapped onto IDL methods. Actions resulting in multiple replies may generate exceptions to draw the attention of the calling object, with the replies modeled as methods in the opposite direction. Notifications may be mapped onto interfaces in the opposite direction, corresponding to the push and pull models. Finally, conditional packages can be made "mandatory" by being added to the resulting IDL interface. Their presence, however, becomes an implementation issue: the standard CORBA *not_implemented* exception should be raised whenever a method of a nonimplemented package is invoked. Translated IDL interfaces follow exactly the same inheritance lattice as the original GDMO classes, while the *Top* class inherits from a *ManagedObject* base interface, which in turn inherits from CORBA's *Object*, as do all IDL interfaces (see Figure 2.15).

The suggested mapping goes a long way toward reconciling the differences of the two object models, but some semantics are inevitably lost in the translation. Most notably, in GDMO conditional packages may or may not be included in an object instance at creation time. This facility allows for the late binding of functionality to that instance, and it may also be used to configure its "mode" of operation. This cannot be achieved through the suggested translation. Furthermore, some conditional packages for the same class may be mutually exclusive; this again cannot be modeled in IDL. If ISO and ITU-T are to adopt the proposed translation guidelines by XoJIDM, they should also instruct GDMO information modeling working groups to avoid the use of conditional packages in a non-IDL-compatible fashion.

A more important difference concerning the translation has to do with the access methods. The operational model of CORBA is that of a single distributed object, accessed in a location transparent fashion. In OSI management, managed objects can be accessed collectively through the CMIS/P scoping and filtering facilities. These may be used for discovery services, for example, "which calls are currently established through that element," and they minimize the management traffic incurred on the managed network. In addition, the same operation may be performed on many managed objects. This not only is an engineering-level optimization but also allows a higher level of abstraction to be provided to managing functions. Discovery facilities may be provided through traders in CORBA, as discussed in section 2.3.3, but the efficiency of such mechanisms, with potentially thousands of transient managed objects in network elements, needs to be evaluated. In addition, the CMIS/P operational paradigm with potentially multiple operations expressed through a single request is lost, unless similar facilities are provided through special CORBA servers, as discussed in section 2.6.

2.6. SUMMARY AND THE FUTURE

Here we summarize the key aspects of the three frameworks, make final comments on their suitability for telecommunications network/service management, and look at possible future directions.

The Internet management framework was conceived mainly for LAN/MAN management. It is a communication framework based on the manager-agent model whose design decisions opted for agent simplicity, shifting sophistication and complexity to manager applications. It has adopted a connectionless unreliable transport mechanism, a rudimentary object-based information model, and a polling-based model for element management. It follows a "lowest common denominator" approach to management standardization, addressing only the absolutely necessary aspects. Version 1 offers little security, while managed object creation/deletion is problematic. Its simplicity for managed devices has made it successful in the Internet network element market. Version 2 has only recently been completed. It fixes some of the problems of version 1 (e.g., object creation/deletion), but its adoption is questioned given the overall cost of the transition compared to the new features. In addition, the security framework has not yet been fully agreed, and it is optional, which implies that the overall framework will continue to be used for mostly monitoring rather than intrusive management.

Telecommunications environments guarantee quality of service. They need to support a high degree of availability and fault-free operation and are inherently connection-oriented. Internet management does not well match requirements such as timely reaction to network events, minimization of management traffic, geographic dispersion of control through distribution, and strong security guarantees. In addition, the simplicity of agents is not a big issue for telecommunications network elements, which are typically complex and sophisticated (e.g. exchanges, ATM switches, SDH Add-Drop Multiplexors). In fact, recent advances in inexpensive memory and processing capabilities suggest that the argument of simplicity is no longer valid. In summary, the Internet management framework is not well suited for telecommunications network management. Network elements with Internet management interfaces can be adapted to OSI/TMN by using the IIMC and to CORBA/TINA by using the JIDM solutions.

The OSI management framework was conceived mainly for WAN management and telecommunications environments. It is a communication framework based on the manager-agent model but has opted for sophisticated facilities in agents, as necessitated by the needs of such environments. It has adopted connection-oriented reliable transport and a fully object-oriented information model. OSI agents offer optimized multiple object access and sophisticated event management facilities that provide expressive power and minimize management traffic. The whole framework follows a "large common denominator" approach to management standardization, promoting a common style for management tasks through the system management functions which address reusability and genericity. Its object-oriented nature has led to object-oriented development environments that provide platform facilities similar to those of OMG CORBA. The key difference, however, is that the relevant APIs are not "standard," and this means that there is no application portability across different software platforms. The sophistication and complexity of the overall framework has delayed its adoption, but early research efforts [32] and recent platform products have accelerated the development process, and there are now a number of elements with OSI/TMN compliant interfaces in the marketplace. In summary, OSI management is ideally suited for telecommunications *network* management, and it has been adopted as the base technology for the TMN.

OMG CORBA has evolved from the distributed system world and can be seen as a pragmatic solution that conforms to the spirit of the ISO/ITU-T ODP standards. It projects a single distributed object paradigm, accessed transparently through the ORB, as opposed to the object cluster approach of the manager-agent model. In addition, it is mostly an object-oriented distributed programmatic framework that standardizes APIs to the underlying ubiquitous software infrastructure, the "distributed processing environment." Its object-model is fully object-oriented and largely compatible with that of OSI management. Underlying communications are reliable connection-oriented approaches, with connection management taken care of by the DPE. The event management model is simpler and less powerful than that of OSI management. Multiple object selection and discovery facilities may be supported through name servers and traders, albeit with increased traffic and reduced timeliness compared to OSI management. The use of CORBA for telecommunications network management is theoretically possible but has not yet been attempted in practice. Federation issues (e.g., for trading) have not yet fully worked out, while the feasibility and cost of an ORB, name server, and trader for every network element with potentially thousands of managed CORBA objects needs to be assessed. The real strength of CORBA is distributed system building, and, as such, it makes it an ideal candidate for distributed service operation and management in the context of new advanced services. In summary, OMG CORBA is best suited for telecommunications distributed *service* operation and management, and this is why it has been adopted by TINA as the basis of the TINA DPE.

OSI management and CORBA will have to coexist in the context of telecommunications network and service management in the years to come. We will finalize this chapter with two scenarios for their coexistence and integration: a pragmatic approach, which takes into account the past and present investment in this area, and a "blank paper" approach, potentially suitable for the long term.

TMN addresses mostly network operation and management, while service management has not been addressed yet by the relevant standards groups. TMN service management addresses mostly "traditional" telecommunications services *not* supported by distributed applications — for example, leased line on demand with guaranteed quality characteristics. While TMN mechanisms can be used to manage advanced services supported by distributed applications (e.g., video-conferencing, joint document editing), this will result in different mechanisms for service operation and service management, as already explained. The TINA framework tries to unify service operation and management mechanisms through a CORBA-based DPE, and it is likely this will be the way advanced telecommunication services will be offered in the future. Despite the fact that the TINA framework intends to replace completely the TMN in the long term, the most likely scenario is that TMN will be used for *network* management and *traditional service* management, while TINA will be used for *advanced service* operation and management.

The reusability of TMN management services is shown in Figure 2.16. The TMN in the left part of the figure supports both network and traditional service management. Its management services are accessible through OSI management-based X interfaces. These services could be accessed from CORBA-based environments (e.g., in customer premises networks), through suitable adapters according to the JIDM specifications. Note that adapters for other technologies, such as the World Wide Web (WWW), may be necessary. The TMN in the right part of the picture supports only network management functionality. TINA advanced services operate on top of it and reuse the supported network management services (e.g., fault/configuration management, network quality of service management). Again, TMN services are accessed through JIDM-compliant CORBA to OSI management adapters.

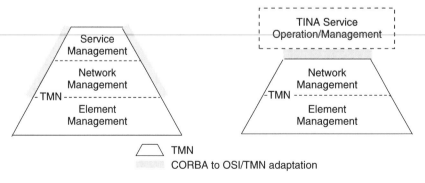

Figure 2.16 CORBA to OSI/TMN Adaptation Scenario.

The second scenario for their integration examines the possibility of combining the relative strengths of both technologies by providing OSI management facilities in a CORBA environment. An ISO/ITU-T initiative that studies the impact of ODP on OSI management is known as the Open Distributed Management Architecture (ODMA) [35]. This is a theoretical high-level study. In addition, research work by the author, described in [36] and by others has specified OSI-SM facilities over CORBA. The approach is depicted in Figure 2.17.

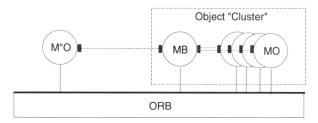

Figure 2.17 The OSI/TMN Operational Model over CORBA.

M"O: Managing Object
MO: Managed Object
MB: Management Broker

In this approach, the operational framework of OSI management is retained over CORBA through Management Brokers (MBs). Existing GDMO information viewpoint specifications are translated to IDL computational ones on a one-to-one basis, using the XoJIDM [34] approach. Managed objects are implemented as equivalent CORBA interfaces with a logically bound cluster of managed objects, similar to an OSI/TMN agent, administered by a Management Broker. The MB provides multiple-object access facilities through scoping and filtering. In addition, it acts as an object factory, naming, and notification server. Event management is provided by event forwarding discriminators and logs, with relevant filter attributes supporting the fine-grain control of notifications. The rest of the OSI SMFs are maintained as generic CORBA objects that may be instantiated within a cluster. In summary, the only necessary CORBA service is naming in order to address the MBs in a location-transparent fashion.

This approach essentially maintains the OSI/TMN operational model over CORBA but replaces the access mechanism (i.e., CMIS/P) through CORBA interactions and the distribution mechanism (i.e., OSI Directory) through the CORBA naming service. Of course, interoperability protocols other than IIOP will be necessary to support interoperability in telecommunications environments; relevant mappings are expected to be produced by OMG in the future. The proposed approach retains the OSI management expressive power, event model, and generic management facilities, while it benefits from the distribution, portability, and easy programmability of CORBA. Such an approach will make possible the eventual migration toward a single integrated "service engineering" framework that will encompass both service and network management aspects.

Acknowledgments

Sections 2.2.1 and 2.4.3 are based on earlier versions by Graham Knight of UCL, from a joint unpublished document on the comparison of the Internet and OSI system management frameworks.

The research work for this chapter was undertaken in the context of the ACTS VITAL and REFORM projects and the RACE ICM project. The ACTS and RACE programs are partially funded by the Commission of the European Union.

References

[1] ITU-T Rec. X.701, *Information Technology—Open Systems Interconnection—Systems Management Overview*, 1992.

[2] J. Case, M. Fedor, M. Schoffstall and J. Davin, *A Simple Network Management Protocol (SNMP)*, RFC 1157, 1990.

[3] J. Case, K. McCloghrie, M. Rose and S. Waldbusser, *Protocol Operations and Transport Mappings for Version 2 of the Simple Network Management Protocol (SNMPv2)*, RFCs 1905/1906, 1996.

[4] ITU-T Rec. X.901, *Information Technology—Open Distributed Processing—Basic Reference Model of Open Distributed Processing—Part 1: Overview*, 1993.

[5] OMG, *The Common Object Request Broker Architecture and Specification (CORBA)*, Version 2.0, 1995.

[6] ITU-T Rec. M.3010, *Principles for a Telecommunications Management Network (TMN)*, Study Group IV, 1996.

[7] *An Overview of the Telecommunications Information Networking Architecture (TINA)*, TINA'95 Conference, Melbourne, Australia, 1995.

[8] ITU-T Rec. X.208, *Specification of Abstract Syntax Notation One (ASN.1)*, 1988.

[9] K. McCloghrie and M. Rose, *Structure and Identification of Management Information for TCP/IP-based Internets*, RFC1155, 1990.

[10] J. Case, K. McCloghrie, M. Rose and S. Waldbusser, *Structure of Management Information for Version 2 of the Simple Network Management Protocol (SNMPv2)*, RFC1902, 1996.

[11] S. Waldbusser, *Remote Network Monitoring MIB (RMON)*, RFC1271, 1991.

[12] ITU-T Rec. X.701, *Information Technology—Open Systems Interconnection—Structure of Management Information—Management Information Model (MIM)*, 1991.

[13] ITU-T Rec. X.701, *Information Technology—Open Systems Interconnection—Structure of Management Information: Definition of Management Information (DMI)*, 1992.

[14] ITU-T Rec. X.701, *Information Technology—Open Systems Interconnection—Structure of Management Information: Guidelines for the Definition of Managed Objects (GDMO)*, 1992.

[15] OMG, *Specification of the Interface Definition Language (IDL)*, CORBA Version 2.0, 1995.

[16] OMG, *Common Object Services Specification (COSS)—Event, Life-Cycle, Name, etc.*, 1994.

[17] ITU-T Rec. X.500, *Information Technology—Open Systems Interconnection—The Directory: Overview of Concepts, Models and Service*, 1988.

[18] ITU-T Rec. X.750, *Information Technology—Open Systems Interconnection—Systems Management—Management Knowledge Management Function*, 1995.

[19] ITU-T Rec. X.710/711, *Information Technology—Open Systems Interconnection—Common Management Information Service Definition and Protocol Specification (CMIS/P) Version 2*, 1991.

[20] ITU-T Rec. Q.811/812, *Specifications of Signaling System No. 7—Q3 Interface—Lower and Upper Layer Protocol Profiles for the Q3 Interface*, 1993.

[21] ITU-T Rec. X.734/735, *Information Technology—Open Systems Interconnection—Systems Management—Event Management and Log Control Functions*, 1992.

[22] OMG, *General Inter-Operability Protocol*, CORBA Version 2.0, 1995.

[23] OMG, *Internet Inter-Operability Protocol*, CORBA Version 2.0, 1995.

[24] ITU-T Draft Rec. X.9tr, *Information Technology—Open Distributed Processing—ODP Trading Function*, 1994.

[25] ITU-T Rec. X.730-750, *Information Technology—Open Systems Interconnection—Systems Management Functions*.

[26] K. McCloghrie and G. Waters, *Administrative Infrastructure and User-based Security Model for Version 2 of the Simple Network Management Protocol (SNMPv2)*, RFCs 1909/1910, 1996.

[27] ITU-T Rec. X.830-833, *Information Technology—Open Systems Interconnection—Security, Generic Upper Layer Security (GULS)*, 1996.

[28] OMG, *Security Specification and Common Secure Interoperability—Version 1.0*, 1996.

[29] S. Bhatti, K. McCarthy, G. Knight and G. Pavlou, "Secure Management Information Exchange", *Journal of Network and System Management*, Vol. 4, No. 3, pp. 251–257, Plenum Publishing, 1996.

[30] X/Open, *OSI-Abstract-Data Manipulation (XOM) and Management Protocols (XMP) Specification*, 1992.

[31] NMF, *ISO/ITU-T Internet Management Coexistence (IIMC)—Translation of Internet MIBs to ISO/ITU-T GDMO MIBs and ISO/ITU-T to Internet Management Proxy*, Forum 026 and 028, 1993.

[32] G. Pavlou, G. Knight, K. McCarthy and S. Bhatti, "The OSIMIS Platform: Making OSI Management Simple", *in Integrated Network Management* IV, eds. A. S. Sethi, Y. Raynaud and F. Faure-Vincent, pp. 480–493, Chapman & Hall, London, 1995.

[33] G. Pavlou, K. McCarthy, S. Bhatti and N. DeSouza, "Exploiting the Power of OSI Management in the Control of SNMP-Capable Resources Using Application-Level Gateways", in *Integrated Network Management* IV, eds. A. S. Sethi, Y. Raynaud and F. Faure-Vincent, pp. 440–453, Chapman & Hall, London, 1995.

[34] X/Open / NMF, *Joint Inter-Domain Management (JIDM) Specifications— SNMP SMI to CORBA IDL, ASN.1/GDMO to CORBA IDL and IDL to GDMO/ASN.1 translations*, 1994.

[35] ITU-T Draft Rec. X.703, *Open Distributed Management Architecture*, 1995.

[36] G. Pavlou, *From Protocol-based to Distributed Object-based Management Architectures*, Proceedings of the IFIP/IEEE International Workshop on Distributed Systems: Operations and Management, Sydney, Australia, 1997.

George Pauthner
Alcatel Telecom
and
Jerry Power
Alcatel Telecom

Chapter 3

Management Platforms

3.1. INTRODUCTION

The increasing complexity of today's SONET network elements and the topologies they support require a different network management solution than the traditional monolithic systems of the 1970s or the technology-specific solutions of the 1980s. These new-age management systems must allow quick adoption of operational infrastructure changes to increase the speed with which new services can be introduced to a network. Far too often, network operators spend a lot of time and money deploying the latest network technologies, only to discover that the means to manage the technology is either nonexistent or lagging far behind. Often, only after the technology is deployed can we determine that the new technology cannot easily integrate with the existing network or that the operational investment required outweighs expected network benefit. Forward-looking software management platforms (sometimes referred to as frameworks) seek to provide a solution to this ongoing conundrum.

What is a software platform? For a political party, a platform consists of the defining beliefs that bind a group of people into a consolidated force. It is described by the tenets that are common to all its members. Within a specific political party there may be multiple coalitions and factions with goals and objectives that are more specific than those of the complete political party. A single political party may host several such splinter groups with goals that conflict with other splinter groups or that complement other factions residing in that political party. Software platforms pre-

111

sent an analogous scenario. Like a political party's platform, a software platform consolidates and manages common functions that are used by independent applications.

A software platform is composed of services and behaviors; it provides a common set of functions for the applications that make use of it. Software platforms can be broad and wide reaching or narrow and focused. A general-purpose network management platform will attempt to provide functional services that are common across SNMP, CMISE, TL1, and perhaps ASCII management domains. A network management platform that is targeted toward providing management support for TL1 network elements in a network operator environment would be a very different type of platform. The general-purpose management platform would have very broad appeal when compared to the TL1 management platform. However, the number of software functions that can be made common across all the targeted management domains would be much smaller than the functions of the TL1 management platform.

The need to support rapid development and introduction of management software to an operational environment drives several key concepts into the software platform environment. Many current management applications contain some degree of replicated functionality. Such duplicated software functions extend development schedules, increase software costs, and decrease the usability by forcing operators to learn multiple systems that achieve a similar end. A reduction application development cost can be achieved when software does not have to be redesigned, developed, and tested for functions that are common across a platform community. In the case of platform development efforts, reduced software cost is achieved in concert with increased functional capabilities since this allows the application development team to focus on solving a business problem and not on solving prerequisite infrastructure issues. Perhaps even more importantly, platform software that can be tested and certified in a variety of applications environments can lead to a higher quality system and one that is easier to support since the support resources can be focused on more common software components. This later aspect often goes unaccounted, even though these long-term life-cycle savings can account for more than 75 percent of a project's software life-cycle costs. A properly segmented and modularized management software architecture can begin to address these issues.

Although a modularized management software design has the beginnings of a platform, it is not a platform system onto itself. The software platform includes all the common software functional modules that are shared between disparate management applications but does not include the application-specific functionality. The interfaces to the platform must be well defined so that independent software development teams can develop software that makes use of the platform services without knowledge of how those services are achieved.

True software platforms exclude any application-specific processes so that an independent integrator is able to obtain applications from multiple sources that run on top of the platform. This allows the applications to be obtained on a competitive basis from suppliers with core competencies to deliver that software. The more complex the communications network, the more important it becomes that the operator be able to integrate multisourced best-of-breed software modules into a single unifying platform architecture. More advanced systems will include the ability for a network operator to replace standard software modules with customized software modules that provide a unique operational personality to the network. As the network environment moves to a competitive arena, such capabilities will be required to allow a network operator to differentiate their service offering from the competition.

Platform benefits can be measured in terms of how much time they save in code development. Such savings will reduce the software price and possibly increase the management functionality. Therefore, the better the platform, the quicker an application developer can build a specific management application. The value of a platform system is directly proportional to the amount of development time saved by the application. An additional advantage achieved from a platform-based management system is that it affords the possibility of achieving a common look and feel to the management of disparate network elements. Each tier in a tiered platform structure presents an interface to the next tier. These tiers provide portals of opportunity to customize the services provided to the software that builds on it. Each individual tier should simplify the task assigned to the tier above and make it easier for that layer to complete its tasked functions. This layered simplification process makes it quicker and easier to make procedural modifications at that layer since the complexity of the software below are masked from its purview. Figure 3.1 illustrates how common functions can be shared between applications.

Historically, network management systems were implemented to automate functions that were manual. Thus, management systems were tools that a network operator used to reduce existing operational costs. Large management support systems were developed by a single company and sold to network operators with similar operational structures, behaviors, and requirements. This continues to be the major driving factor behind the evolution of most nonplatform management systems. Over time, such systems achieved a degree of flexibility through the consolidation of requirements from multiple network operators. However, each served network operator ends up bearing the weight of the other network operators' peculiar

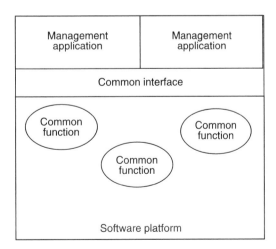

Figure 3.1 Platform-based Applications Share of Access to Common Functions.

requirements. Unfortunately, this evolution tends to lead to functional isolation and serves to further lock in historical practices that will prove untenable as markets move to a more competitive environment.

A single computer can have multiple nested platforms. Platforms can be layered, and not all applications need to access the platform at the same layer. More importantly, if a special-purpose platform is built with the expectation that a more general-purpose platform will be running in support of its needs, the special-purpose platform can take advantage of the situation and build up from where the general-purpose platform ends. Tiered platform systems are often called middleware. In Figure 3.2 the middleware platform extends the general purpose platform functionality for the management applications with common requirements.

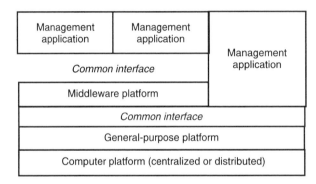

Figure 3.2 Middleware Extends Platform Functionality.

The Telecommunications Management Network (TMN) architecture was an attempt to use standards in an effort to break free of archaic operations dogma. TMN attempts to drive the operations support systems to a universal object-oriented communications mechanism and away from regional standards such as TL1. TMN provides a structural base that can be used to describe and implement a management system. The goal of TMN is not to constrain the development of network element features. It was never intended to force a carrier into a specific procedural flow. Although this should seem intuitive, it is a concept that is often missed or misconstrued. A true TMN system provides a consistent and efficient mechanism for the collection and storage of management data. It provides an independently flexible application interface to the management data. It also allows the operator to rapidly change the network elements and collection techniques or change the operational procedures used to manage those elements. In a competitive network environment, the network carrier that is able to introduce services quickly, rapidly activate customers, change operational procedures, and provide timely network element upgrades will be the network operator that survives. Although TMN does not require the use of a software platform, software platforms have many of the same goals as TMN and provide the technical keys needed to achieve the modular independence envisioned by the TMN framers.

There are many different kinds of platforms. A platform can be management specific, or it can provide service of value to nonmanagement applications. Management platforms can provide a complete functional TMN layer or a portion of a TMN layer. Platforms can build on other platforms. They can also provide functions with specific purposes, or they can be general purpose and used by many associated processes. Platform software is often referred to as middleware because it is located between other software processes. The services provided by the platform vary widely and can be modularized into functional areas. An advanced and more targeted management platform can be built at a systems level by integrating middleware and base functional platforms from many sources.

Middleware is often used to provide the client-server connectivity software that supports the enabling services which allow multiple processes running on one or more computing engines to interact across a network. Network management is inherently a distributed activity that has historically been artificially forced into a rigid centralized topology. Client-server software concepts are idly suited to take advantage of modularized platform software. In the client-server paradigm, common services can be distributed to distant computers in an effort to manage processing loads. Different connectivity platform software may provide support for the Open Software Foundation's (Cambridge, Mass.) Distributed Computing Environment (DCE), the Object Management Group's (Framingham, Mass.) Common Object Request Broker Architecture (CORBA), and Microsoft's (Redmond, Wash.) Object Linking and Embedding (OLE). Such software provides the framework that allows processes to remotely link into the interface supporting the common processes of the software platform. A distributed systems platform may make use of one or more of these connectivity software packages to achieve the scalability needed to enable a management solution to grow from initial deployment to large-scale systems. Historically, management systems had to be engineered for the largest envisioned deployment scenario upon initial deployment—client-server architectures break this paradigm.

Communications services are also key candidates for bundling into a support platform. These services provide support for synchronous, asynchronous, transactional, and interactive communications between management application processes and between management applications and the network elements they support. At the top of the communications layer, application management platform middleware may offer directory, security, data access, and other management services.

Data management services are easily encapsulated into a middleware software layer. They may provide Management Information Base (MIB) systems management, object persistence, and database systems support. From a software process perspective, the supported network elements appear as a persistent source of data storage. Network operations can be affected by modifying attribute values, reading operational status registers, and creating or deleting network resources. Network-level data are manageable in a similar fashion. Such data services can be used extensively by almost all manager and agent processes to effect changes in the network.

User presentation services are often neglected by telecommunications standards activities. However, they are extremely important if the network operator ever hopes to achieve a common look and feel across divergent manager processes. Although effective user-interface standards are essentially nonexistent, many examples of presentation service vehicles are available in the computing industry. Presentation services may range from map management processes to widget libraries that simplify graphic icon constructs for the user.

3.2. THE WELL-DESIGNED PLATFORM SYSTEM

A well-designed management platform starts with a framework that defines a system architecture that manages the interaction between system-level components. A distributed platform that is scaleable eliminates the physical constraints that impede the growth of the legacy management systems. They allow administrators to develop a management system without taking the system off-line, and they allow the cost of the components associated with the larger systems to be deferred until growth is needed. Distributed systems also permit load balancing over time to achieve a balanced processor loading to optimize system performance. As the demands placed on the processor fabric change, management and agent processes can be rebalanced to fit the changing environment.

Management applications run on computers far removed from the devices that operate on the network and are the source of status data. In the days of large mainframe computers, applications were limited by the size of the host computer. The managed network devices were simple and relatively easy to manage. Network management was largely devoted to remote detection of warning conditions and then central coordination of the manual minions that isolate and correct the situations. The mainframe approach has given way to distributed processing techniques that programmatically link multiple computers into a cohesive computing fabric. Such a computing environment eliminates the bottlenecks associated with a centralized system. Rather than having to size their management system for the ultimate growth scenario, administrators can start with a smaller computer and grow their environment by introducing additional computer engines. As computer technologies change, the new computers can be linked to the older systems in order to support a graceful migration of the support system. Functions that were never considered at initial deployment can be gracefully added to an evolving systems architecture; migration of management systems in an incremental fashion then becomes possible.

Most modern management applications control the network using object-oriented techniques and interact with platform middleware services via RPCs (Remote Procedure Calls). RPCs can be used within a single machine as a supplier line of demarcation or across machine boundaries as a client-server deployment boundary. As shown in Figure 3.3, RPCs allow an application to call a procedure on a remote machine as though it were part of the same application on the same machine. The RPC call establishes communications and handles data translations for the user.

Platforms provide a degree of uniformity that allows software components from multiple sources to be plugged into a common infrastructure. These software components from multiple sources of origin all reuse the services of the software plat-

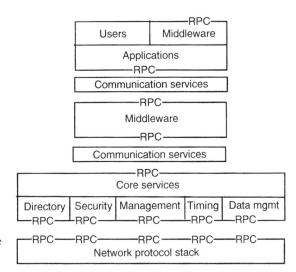

Figure 3.3 Platform Interface Linkages.

form to reduce the complexity of their given task. Individual software modules may evolve independently from the software platform, just as the software platform evolves independently from the components that use its services. Thus, platforms help support the structured evolution of the management environment. Platforms can standardize the systems protocols, the operations performed against the MIB, and the models maintained by the MIB.

The same forces that are driving the movement to distributed computer systems are also increasing the intelligence resident in the network elements. Network elements are becoming increasingly capable of more complicated management tasks, their capacity is larger, and they are becoming more flexible in terms of functional deployment. Growth in network element intelligence permits more intelligent services to be offered to the end-user, many of which have sophisticated management systems for management of the Customer Premises Equipment (CPE) which they maintain. It is easy to see the recent growth in service support applications and to measure the strain these systems have put on management systems that were not designed to handle the load. This trend shows no sign of diminishing in the future and, as such, needs to be accommodated by system design. Such unprecedented growth in the demands for management processing is exponentially increasing the rate at which embedded support systems are becoming obsolete.

In a single network, network elements come from different vendors depending on which vendor is offering which innovative technologies and at what price. Equipment vendors compete against each other, and this competition keeps costs down and innovation high. Software is no different. Platforms need to be open, well defined, and flexible so that different software houses can compete to supply innovative software components. Such a systems-level concept requires defined interfaces and functionality, not just between computers but between software components within that computer. Such defined interfaces are called Application Program Interfaces (APIs). In order to encourage the innovation demanded in a competitive network operator environment, the API definitions should be functionally specific but open to enhancement and evolution.

Competing carriers purchase their network elements and management platforms from the same pool of network element suppliers and have achieved a degree of service differentiation through their engineering practices. As networks become more dynamic, engineering will be transitory and the mechanism for differentiation will be through operational procedures. Thus, a key requirement for a management platform will be the ability of a carrier to modify operational procedures within the system. A carrier's strategic plan will need to be tightly coupled to its operational procedures, and carriers will need to be able to adjust their operational policies rapidly in a competitive environment. Operational policy definition and policy management are a function that should be supported by the management system and not constrained by a regiment's structure and implementation.

It is important that a TMN platform eliminate duplicated data points within the network and facilitate the sharing of data between applications. When data are duplicated and exist in two places within the management infrastructure, it becomes difficult to ensure that the data sources stay synchronized. Should an error be injected into the synchronization system (these are unavoidable), and one value change, system resources must be able to determine which value is accurate once the synchronization system comes back on-line. Management of duplicated data is the Achilles heel of many systems because of its reliance on human intervention to resolve and manage data duplication issues. Whenever a management application needs a data point, no matter which of many applications needs that data, it is important that a consistent value be accessed.

TMN architectural rules divide the network management function into four layers—the business, service, network, and element layer. The Element Management Layer (EML) span of control will vary from a single wire center to multiple wire centers depending on the concentration of traffic and network elements supported in that geographic region. The EML acts as a front-end processor that provides a logical view of the supported network elements to the network layer. This logical abstraction frees the network management layer from having to know all the specific physical aspects associated with the network element supported by the EML. The EML filters events from the network elements, manages the connectivity within its span of control, and supports the equipment-specific processes needed to consolidate network elements.

The Network Management Layer (NML) uses the functionality of the EML to provide a higher degree of functionality across an entire network. The higher the degree of abstraction that can be provided at the EML, the simpler the NML because more of the cross-domain management functionality has been distributed further into the management structure. Since the NML builds on EML functionality, from a NML-centric perspective, the NML can view the entire EML as a base network platform that it builds upon. In the highest state of TMN evolution, there is very little functional difference between the EML and NML—the difference is in the breadth of their network purview. Since EMLs can themselves be built from platform systems, a recursive platform relationship exists between the network at these layers. As EMLs evolve, they will become viewable as objects in their own right and will be managed by the NML just as the current generation of EMLs manages the network elements.

The EML's ability to support the required functions are dependent on its breadth of control. For example, the more network elements supported by an EML, the more knowledgeable its root cause analysis (RCA) rule base and the better able it is to filter events to the NML. EMLs that are technology-specific are unable to see the interaction between technologies and unable to support evolving network elements as they integrate technologies into a single system. For example, a light-wave-only EML would not see how the actions of a cross connect affect the alarm patterns reported by the lightwave device. In addition, an Asynchronous Transfer Mode (ATM)–only EML would not be able to manage a lightwave device that had an integrated ATM concentration feature. Successful deployment of such an EML system dictates deployment of an open platform-based technology that is capable of supporting a multivendor and a multitechnology environment.

3.3. METHODS AND TOOLS

Platforms can be layered in a series of functional entities. Different layers provide different services to the processes that build upon their capabilities. The primary network management platform layers include the Guidelines for the Definition of Managed Objects (GDMO) toolset, MIB management and data storage tools, the Graphic User Interface (GUI) presentation platform, process management, communications infrastructure, protocol stack support tools, common interworking applications, and application framework tools. APIs are ordinarily used to link functionality across layer boundaries. These layers are illustrated in Figure 3.4.

3.3.1. GDMO Toolset

Platforms are vehicles that allow software from many sources to come together quickly and function as an integrated entity. The platform itself may be composed of software from multiple sources. The successful integration of all this software requires a mechanism to define how these processes link together. Most modern-day management systems are based on object-oriented techniques. Object-oriented systems model the resources supported by the network elements as a series of objects, each of which supports a series of attributes. When a manager application wishes to change an attribute of an object, it sends a command to a proxy agent and the agent changes that operational parameter in the network element. To extend the paradigm even further, the manager applications are themselves software objects. A higher order management system can change an attribute of a management process by sending a command to that process. If a process knows the object definition that describes the resource it wants to manipulate, then it can manage that object.

Common Management Information Service (CMIS) is a language that allows a manager to send object-oriented commands from a manager to an agent to manipulate objects. CMIS commands reference an object that is defined in a GDMO

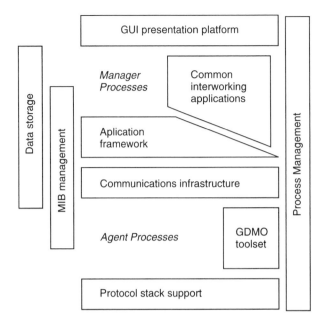

Figure 3.4 Platform Components.

(Guidelines for Defined Managed Objects) description. GDMO is a formalized specification language that allows a user to describe an object and its behaviors in an information model. If one were to think of the network as a series of objects that provides a set of services to the management system, the GDMO would define those services and how to use them. Although standardized GDMO templates can be used to reduce the labor required to model a network element's behavior, they cannot provide 100 percent functionality for constantly evolving network elements. The amount of manual labor involved is proportional to the amount of value-added features the network element provides over the standard model.

GDMO tools simplify the process of developing agent and manager processes. When a developer is building a process that will manage objects with a defined GDMO, the developer can load a GDMO description into the GDMO compiler, and the tools will produce a structural outline of the software needed to manipulate that object. This saves the developer from the labor-intensive task of hand-coding the interface software.

GDMO language is extremely complicated, but many generic GDMO compilers accept GDMO object descriptions and generate C++ code and data structures that can be used as a basis for agent or manager development. GDMO compilers often produce code stubs that support object filtering, scoping, persistence, object creation, and object deletion services. Many platforms include editors specifically designed to aid the developer in manually editing complex GDMO descriptors. When the user is satisfied with the results, the descriptor can be fed into the GDMO compiler.

Other GDMO tools also exist to provide a greater level of platform-specific support. For example, while the generic GDMO compiler assumes the existence of a CMISE network element, some GDMO toolsets allow the user to characterize a TL1 network element and uses this characterization to build the GDMO support code. These platform-specific toolsets also generate linkages to meta-data services, TL1 communications services, and manager applications to the managed object class level. Although some efforts are underway to automate as much of the agent-manager construction process as possible, the process will always involve some degree of manual intervention that is needed to account for vendor differences in network element features, organization, and structure.

Many GDMO compilers are able to use the GDMO descriptions to load MIB browsers. MIB browsers are applications that allow the user to traverse the resource tree and examine the objects that are contained in the MIB. Many platforms provide generic tools in order to provide a graphical user interface (GUI) that can be used to interact with and manipulate the managed objects contained in a network element. An import and export tool that accepts GDMO object definitions, imports them into an object dictionary, and dumps the GDMO descriptions from the loaded dictionary (this includes a semantics checker with a report generator) is provided. As an open system, network operators can select and use any GDMO browser they desire (each allowing the other to fully browse any object defined to the system).

Object-oriented techniques were developed to support layering of the abstraction process. The more abstraction, the simpler to perform complicated management tasks. Some GDMO-based tools are beginning to appear which provide more sophisticated and more capable interfaces to the platform superstructure. The Object Management Group (OMG) Common Object Request Broker Architecture (CORBA) interfaces are an example of such next-generation interface systems.

CORBA-based Interface Definition Language (IDL) interfaces are a more flexible alternative to the CMIS interface. They make it easy and efficient to manipulate a manager application that is in and of itself a composite object. However, they were designed specifically for process-to-process communications and are triggered by an RPC between two cooperating processes. IDLs can be hard coded into an application, linked into the application, or dynamically loaded. Hard-coded IDLs are functional for interfaces that do not change; they are fixed and do not allow easy migration, but they do provide a high-performance interface option. Linked IDLs are accessed through coded libraries and can easily be changed through the use of a toolset designed for that purpose. Dynamically loaded IDLs are sometimes call contracts that can be registered with a server. The manager application obtains the IDL and agent addressing information from a server before attempting to send commands to that agent. Dynamically loaded APIs have the most value when manager or agent applications do not have a defined residence and can float through the distributed computing environment.

3.3.2. MIB Data Management

GDMO toolkits build the key components of the infrastructure that supports the management network. One of the key components that underlies an object-oriented management system is the MIB. The MIB contains the collection of managed object classes. The object class definitions include a list of attributes, operations, actions, a description of its relationship with other object classes, and events associated with the managed network elements. For each network resource that can be managed or monitored, the managed object class contains a managed object. As new objects are added to the network, new managed objects are instantiated within the MIB, and for each instance of a managed object class there is a one-to-one relationship with a physical network resource. The managed objects are an abstraction of the real managed resources that are contained in the network. For example, the managed object description for a NE contains the following:

- Alarms that affect the object's state.
- A description of the attributes that affect its provisioned function.
- Actions needed to effect a cross connect in the network element.

The object classes are organized into a containment hierarchy referred to as a containment tree. A single network element contains several managed object classes, and the containment process allows the operator to manage groups of objects that are scoped together under a higher order managed object. Each object in the containment tree has a fully distinguished name that provides an absolute unique naming reference for that object. Objects also have a relative distinguished name that uniquely identifies the object within a local area domain of the containment tree. The registration tree is used to provide object identifiers and location that reference objects by physical location. With the object identifier, the management network can create objects, delete objects, get object descriptors, and set object attributes.

Thus, every time the network changes, the MIB will change to reflect the evolved network status. A single change may have a ripple effect across a series of related objects within the MIB. For example, when a DS3 goes into alarm, the managed object that represents the facility will change its state, the card that supports that facility will change its state, and the shelf and NE superstructure that support that card will also change their state so that the user can see the effect of the alarm through the containment tree. When operators wish to modify an operational parameter of the network, they produce that change by changing an attribute of the managed object that represents the network resource in question. That change may effect changes in supporting objects and will eventually effect changes in the NE directly.

Once an MIB has been constructed, it must be maintained. Proxy agents interface with and manage the MIB so that it reflects the resources of the managed network element. When the management system wishes to modify a network element, the requests are directed to the proxy agent and the proxy manages those changes on the network element. When the network element changes (due to an alarm or a local craft operator action), it will notify the proxy agent that the proxy agent is responsible for communicating that object change to the rest of the management structure. Proxy agents provide an object-oriented means of managing the network even though the network may not be object oriented. Proxy agents are processes that can run in the network element control system or in the network management system. Network element resident proxy agents directly manipulate the network element's resources in response to management system requests. Management systems use network management resident proxy agents to allow the interface between the management system and the network element to be TL-1 until CMISE interfaces begin to appear in network elements. A properly constructed proxy architecture will allow TL1 network elements to behave as a CMISE network element in order to support the seamless management of these network elements during the transition period. The information models for TL-1 based network elements will be limited by the objects supported by the network elements. The TL-1 models are flat in that there is no relation between entities supported by the TL-1 message structure.

The inventory of each network element can be retrieved from the network element and stored in the MIB. Through autodiscovery a management system does a full inventory retrieval as part of activating a network element and sets the MIB to reflect the current network element status. Once the management system compiles inventory and state information from network element autodiscovery responses, the management system uses the autonomous event indication messages from the network element to keep the MIB synchronized with the network element. The MIBs from the various network elements are then linked together in accordance with the way the network elements are interconnected and maintained as the network is expanded. The process is often referred to as topology management.

The MIB can be divided into sections relating to specific network element types. Users can dynamically load the MIB (or, conversely, unload the MIB) with the GDMO that relates to specific network element types. Such actions effectively turn on or turn off support of a specific network element type. Once the MIB is loaded, the network management applications have complete control over the MIB objects. MIB objects can be modified by MIB browsers or by any other management application.

Not all network information can be autodiscovered from the managed network elements because knowledge of an individual network element is limited to the boundaries of that functional agent. Some network information is used to describe the logical relationships between network elements and the services they support. For example, an object describing a path construct has a status and a series of attributes. Although a path is not a single tangible resource, it points to a series of physical network elements that are involved with the path and to the optic spans that tie them together. In addition, the optic spans are facilities that do not report to the network management system but are involved in the linkages that make up the end-to-end path. The network management MIB has to provide a means to maintain and to manipulate the logical MIB as well as the physical MIB that is derived from the network elements proper.

MIBs can also be mirrored. That is, an object resident on one computer or network element can be structurally replicated on another machine. The management system designates one object as the primary object and the other as the secondary object. Whenever the primary object changes to reflect a change in the managed network element, the management system issues a notice that results in the secondary object changing state accordingly. Such a mechanism allows for object redundancy across a distributed computing environment. When the computer that supports the primary object fails, any object requests can be transparently redirected to the secondary object. When the primary object is restored, the two objects resynchronize and continue normal operations.

Meta-data services provide a mechanism to define, store, modify, and retrieve the MIB data much as would graphically be done from the MIB browser. Applications use meta-data services to interact with existing objects, to instantiate objects, and to delete objects. Meta-data service modules can be accessed via programmatic APIs such as XOM/XMP.

Logical MIB management is best accomplished with some sort of data storage tools such as a database system. In the EML and NML, object-oriented management systems model the network relationships in the MIB. A graphic depiction of the MIB would be structured as a mesh network that shows the complexity of the relationships that exist within the physical network. The MIB data structures are best supported with an object-oriented database that is designed for these complex schemas, and they eliminate the constant conversion between object and relational record references in order to speed performance. Many general-purpose database systems are available for use as a source of persistent storage. Relational database systems are a mature technology that does well handling large record volumes that are related by a limited number of key attributes. Object Oriented DataBases (OODB) represent a relatively recent advance in database technology which is designed to support record structures with complex attribute linkages. OODBs differ greatly from their relational brethren in that they expect complex relationships to exist between objects based on the object definition and not necessarily on attribute contents.

Databases provide a means to persistently store the network descriptive information in a common repository that can be shared across multiple applications. Care should be taken with the deployment of a database in support of a network management system to minimize data replication. The data storage tools should provide the tools to minimize or completely avoid duplication of data that can lead to synchronization issues.

As proxy agents and CMISE network elements report events indicative of a change of attribute status, the MIB management tools will update the MIB so that the user can see the attribute change. This is depicted in Figure 3.5. The data storage tools (for example, an OODB) are used to provide support for those attributes not directly reported by the network element but related to network element objects. For example, the state of an optic sheath would not be reported by the network elements that terminate the fibers within the sheath. However, as the network element reports the state changes of the optic terminations, the state of the optic sheath should be updated.

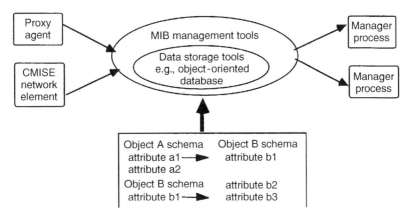

Figure 3.5 Data Storage Platform Components.

Logging services can also provide a data management component of a telecom platform. There are generally two means of retaining log information: save the logging data as a flat file and log records as objects that are stored and browsed in a MIB-type data structure. While the first mechanism is simpler and easier to implement, the second provides a greater degree of search and format control. Such logging data are of critical importance to the network because many complex network issues cannot be resolved without an adequate history of network activities.

3.3.3. GUI Presentation Services

GUI display platforms provide a common user interface for access to manager application processes. Operators access manager applications through menu bars or through icon objects. Once the user has accessed the desired management process and maneuvered to the desired level of functionality, dialog boxes are used to capture request parameters. The graphic applications then use the available APIs to access the platform areas which resolve object names and generate any parameterized action requests for submission to a distributed proxy agent. Such GUI toolkits can reduce development time and software code sizes by as much as 80 percent compared to software written via more conventional processes.

The GUI system is a platform component that is independent of the processes that manipulate the managed network objects. A GUI platform allows multiple application processes from different sources to be accessed from a single display station. These platforms also manage the information display requests and create an integrated display station that depicts status information from multiple sources on a single image. Such system interworking may vary in complexity from X11/ Motif interworking to more complicated information interworking using object interworking or OpenDoc application interworking specifications.

GUI platforms must track an arbitrary series of linked background image displays. They must also be able to maintain a database of which graphic objects are registered on which image display. The database descriptor of these graphic objects may relate the objects to other image displays or to active processes. This graphic management system must also be able to relate icon status to object status as reported by the network. These icons use highlighted status indicators to draw the operator's attention to any outstanding network conditions. Administrators should be able to add new images under an existing image, add new processes to defined image maps, or add symbols (for example, facilities) to the display images.

For flexibility, the graphic platform should not introduce any limits to the number of map levels and should not introduce any procedural prejudices into the management operational process. Maps should be thought of as a graphic containment tree. The objects displayed at one mapping level are contained within the object that references that submap on the parent image. Icons are graphic representations of an object, and the color of the icon is reflective of the object's state. When the status of an object on a submap changes, that status change is visually reflected in the graphic object that contains it on the parent map.

Administrators can actually use the GUI platform to construct multiple views of the same network. Different views may present themselves to the operators differently but link to the same managed objects to ensure that the different operator types have the same perspective on the same network elements. For example, one network operator may have a hierarchical perspective based on the logical organization of the network, another network operator may look at the same objects but their map displays are organized to view the network from its physical structure, and a third network operator may be limited to viewing network exceptions from a report-organized display structure.

A flexible GUI builder toolkit may support a multitude of graphical formats to allow for importing graphic files created in CAD, GIS, and other graphic tools. In addition to the display of graphic screens, network management platforms allow administrators to attach GUI controls (buttons, alarm indicators, gauges, etc.) to graphical screens and associate Fault, Performance, and Configuration Events to these controls. It is also desirable to enable the user to execute configuration commands or queries to network elements via the GUI interface in a customized and context-aware manner. It is also required that the graphical presentation system allow for drill-down capability from one graphical layer to the next.

Network element objects are associated with a network element. On the screen builders, the user picks an icon and places it on the screen. The management application then asks the user to associate that icon with an application associated with a network element class and asks for the information needed to instantiate the network element. An element manager is then initiated, and it causes the proxy to go out and find (then instantiate) all managed objects supported by that particular network element. The element manager application contains all the graphic screens needed to support a user zoom-in on that network element.

Icon libraries and display management platform tools are used to organize management displays. Network operators use these tools to construct maps that depict their network organization and structure. Operators begin with a graphic map of the network and use the icon library to select which icons are located in what area of the network. Icon placement is ordinarily operator controlled in order to prevent one icon from obscuring the display of the status associated with another icon. Intelligent icons, or widgets, are used to manage more complicated tasks.

In an ideal management environment, all operators would use the GUI platform to interface with a similar look-and-feel to the manager and agent processes that support the network. Commonality of presentation allows managers and agent processes to be developed by independent organizations. The operators would use a drop and drag or point and clicks interface to submit commands to agent processes, and the agent processes would change the status display as information is received back from the network. Unfortunately, this ideal has proven very elusive. Luckily, many network managers provide an X-window X11/Motif GUI that supports single-step, point-and-click, mouse-driven operations. This graphics package can capture images, generate graphics, and link graphic objects to MIB-managed object states. Remote X-windows terminals can be connected to the network management platform over any conventional data communications mechanism such as Ethernet or X.25.

While X11/Motif displays increase graphic image portability, they are resource intensive since complete bit-mapped images must be transported across an already busy support infrastructure. Java and other Web-based display technologies are stepping up to provide a solution to this problem. The term *Java* implies an application language targeted at the manipulation of distributed objects and a computer-independent platform for processing commands in that applications environment. Java-based applications (or fragments of a large application often referred to as an applet) can be downloaded from a server to a remotely distributed client environment for local execution. The client computer runs a process known as a virtual machine which accepts and manages the Java applications. Java applications manage the GUI displays, any object retrievals, and modification requests in accordance with the user's demands. From the virtual machine's perspective, it is being served by multiple distributed servers that act as both object brokers and application repositories.

3.3.4. Process Management

The management system is made up of a series of distributed management and agent processes that exchange messages over a computer network. Within the management system, the computers that make up the network and the connections between processes for an internal process network must also be managed. This network must detect and report computer failures, loss of interconnection communications facilities, and individual process failures. Managing of such a network of processes is often as difficult as managing a large network of network elements.

Distributed application processes must communicate with each other to exchange commands and status information. Client-server relationships are normally established between such cooperating processes. The client process is the application that issues a request to the server process. The server process will process the request upon receipt and return a response to the client application process. A single-client application may use a number of server processes to accomplish its desired end goal. A server process may need to communicate with other processes to complete the requested processing, and in doing so it may become a client of some distant server application as well.

Client-server applications are most often thought of as having a two-stage architecture, and this is indeed the simplest case of such a software structure. However, it should be noted that three-tier architectures are becoming very commonplace. In a three-tier client-server architecture, a client application will submit a request to its server for fulfillment. The server application will break the request into a series of simpler requests and submit each request to an information repository that supports that middle architectural layer. Thus, the middle layer software acts as a server for the client software and a client for the information repository. This three-tier client-server architecture is shown in Figure 3.6. Such architectural structures are most often used to avoid the fat-client syndrome that occurs when the size and complexity of the client software become so great that it is unmanageable. Since these structures can be more easily distributed, it can also be used to make the server applications more performant and scaleable. By modularizing and segmenting the server processes, they can be distributed and more easily grown and maintained as well. In extremely large systems (such as a telecommunications management system), the three-tier structure can be easily extended into a multitiered model in which each tier is defined to provide a value-added benefit to the tiers it supports.

To manage a network of processes, each process must be well behaved. Well-behaved processes register their existence with a controlling authority and periodically generate status messages that indicate their well-being. The process manager periodically scans the computers and the processes that have been registered to validate their proper operation. When the process manager uncovers a fault report or finds no report, he or she will notify the system operator and possibly initiate recovery actions.

Many process management systems include performance management tools that look for processes displaying the characteristics of a problematic process, even though the process is not reporting an explicit problem. For example, these systems can look for excessive memory utilization, disk exhaustion indicators, and a host of other symptoms of an underlying problem.

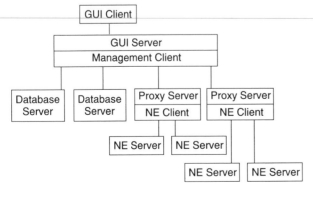

Figure 3.6 A Possible Multitier Client-Server Management Structure.

3.3.5. Distributed Communications Services

The distributed communications services infrastructure includes services to link the computers that make up the network management system. Most modern-day management systems are scalar and are based on client-server concepts. Management and agent applications can be combined on a single server, or they can be distributed through a linked computing environment depending on the specific management requirements. The various managers and agents communicate via an interprocess protocol such as Common Management Information Protocol (CMIP) or CORBA. With the chosen architecture, it is possible at will to implement all components on one server computer or to distribute them on multiple computers.

The distributed communications services define and manage the software objects that are distributed over a network. These services allow a client application to access and use distributed objects to accomplish a defined task. In addition, this structure allows the network operator to gracefully develop the management system as the network grows, as management requirements changes, or as network technology evolves.

In a distributed system, the manager and agent processes can be distributed throughout the network. They can be located in the vicinity on the network elements they support in order to speed performance, or they can be centrally located in order to aid in control of computer resources. The communications infrastructure provides a vehicle that links the processes together; a manager submits an action request to the communications infrastructure, and the system determines the location of the destination process and works to ensure that the command is properly delivered.

The communications infrastructure provides the event-forwarding services that also allow messages to be replicated and delivered to multiple destination processes. This is an important feature for event notification messages since more than one process may be interested in being notified about specific network events.

Event filtering is another service provided by the communications infrastructure. This service limits messages of minimal interest, which reduces the overhead placed on the communication resources used to link the computer systems together as a cohesive fabric.

Management systems can be viewed as a complex web of clients and server processes. Servers are processes that provide a defined function for a requesting client application. A single management application may be a client for one or more application processes and a server for independent application processes. Further complicating the picture, multiple clients may share access to multiple servers, and the server processes may be dynamically assigned by some controlling resource broker. To ensure that the management system can grow without restriction due to geographical or processor bottleneck problems, the client and application processes may be distributed on different and distinct processors. These processors and the client-server processes that communicate between them are linked together using some communications protocols.

Object brokers play an important role in distributed communication platforms. These brokers allow objects and processes to organize themselves dynamically in the management infrastructure. In such an environment, the brokers are used as sources for object locations, provide authentication services, and manage the information flow between cooperating client and servers. Thus, the client applications do not need to be cognizant of the exact physical location of the objects they are accessing. All object accesses are managed through the broker, and this results in a location-transparent client application. Effectively, brokers simplify growth, reliability, and maintenance of a distributed object-oriented system because it allows the objects to be moved without impacting all the accessing client applications.

3.3.6. Protocol Support

The management system must exchange messages with the network elements and between other processes that may be remoted on distant computers. Most network management systems must support network elements from multiple suppliers, and not all these network elements use the same data protocols for exchange of management information. Many more intelligent network elements support a range of data protocol options so that the user can tune the support environment to meet individual requirements.

To support this engineered diversity, the protocol stack support structure should be independent of the manager applications. A management application should have the ability to exchange management information with a lightwave network element over either a LAN/WAN connection (OSI or TCP/IP) or an X.25 connection. This flexibility is achieved by pushing the protocol support structures into a platform layer of software. To change protocol support structures, the network operator simply modifies the software configuration.

At the management level, the most commonly used protocols are Simple Network Management Protocol (SNMP), CMIP, and Transaction Language One (TL1). Each of these management protocols has different strengths and weaknesses.

3.3.7. Interworking Applications

In a network management environment, there are many application processes that can be shared as common across multiple network element types and across multiple functional domains. By pushing these common applications into a platform architecture, a common presentation to the network operator can be achieved and duplicated development efforts can be eliminated.

As an example, the overall platform architecture may include a common alarm management and display system. Any other processes that detect an alert condition requiring operator notification will forward an indicator to the common alarm management system. Alerters are used to bring operator attention to newly received events that require immediate operator attention. The alarm management system tracks open alarms, manages the operator notification and event browsing processes, and logs the network activities. A common alarm management process means that all events from throughout the network will be presented on a single screen that allows the operator to see the relationship between disassociated network events.

Typically, users can use these specialized browsers to filter their displays according to operator-defined criteria. Operator-specified sorting of displayed information on various event keys is also a standard feature of such system. To target operator attention, these systems usually display key attributes of each reported event object; however, a more detailed display of parameterized information is often available.

Test management and performance monitoring applications are another target of implementation as a systemwide interworking application. Performance monitoring applications are in the background and periodically collect performance measurements. These applications usually compare gathered information with operator-established threshold levels, and they sometimes provide statistical forecasting functions that use the compiled performance histories and forecast trend data in an effort to forecast problem areas before they become significant. Test manager applications can make use of topology information to work their way across a defined network path to compile path level test data, or they can work with established test access equipment. Operators can link test access equipment into the network at defined test points to generate detailed traffic profiles. After collecting test information at various test points, the data can be compared to generate a profile of how performance is affected through network sections.

Other systemwide network applications include asset management and capacity-planning tools. Asset management tools collect resource information from the network and manage those resources as financial assets. Resources that are scheduled for future activation, that make up a spares' pool, that have been directed to a repair facility, and that are scheduled for deactivation are all tracked by such a system. Related to these applications are capacity-planning tools that examine the traffic-carrying capacity of the network and use manual projections or statistical use data to project network exhaustion.

3.3.8. Application Framework Tools

Application framework tools seek to reduce the complexity and time required to construct manager applications by automating much of the code used by these processes. Although manager applications will always require some customization to reflect the individual personalities of the network elements supported, much of the processing is actually common across a wide range of management application processes.

Event management services are an example of the tools used to link applications together in a structured environment. Two key components in the event management service chain are the event sieves and the event browsers. Event sieves provide filtering of autonomous event messages. Processes can register with an event controller and indicate the type of events they would like to be informed about. When the event controller receives events they are copied to all processes that have registers as being interested in that type of event. Events that do not fit the registration profile are filtered from the processes that are unconcerned about that type of network activity. Event browsers are GUI-based applications that allow the operator to browse through the event, logging, and archived activity information. Information displays can be organized to fit the operator's individual needs. (Filtering, sorting, and other display functions are supported.)

Scheduling service is another example of a framework tool. Many operations should be handled at a future date on a periodically repeating basis. The scheduling services tool accepts the registration of any such activities and ensures that these processes are initiated at the appointed time.

Help tools are evolving rapidly to increase the depth and breadth of the information available to the operator. Many platforms are basing their help system on standardized structures so that they can be extended to provide greater knowledge to the network operator. Help information is no longer being limited to help on the management systems but is being extended to describe network element specifics as well. It is all being hypertext linked together and interrelated with current network conditions for ease of access in a network critical situation.

The management platform can also provide software administration tools. Software tools are used to manage an environment of network elements that are increasingly software driven. These tools will allow the operator to examine the software versions on the distributed network elements and to remotely update them on an as-needed basis via various software download tools. Many such software packages have configuration and preference settings that are adjustable by the centralized management system.

Trouble-ticketing tools can also be included in the platform infrastructure. Trouble-ticketing system track network faults from inception until closure. Trouble tickets can be used to link multiple events that are related to a single network activity, to manage field resources, and to provide customer quality assurances. The management platform provides the linkages to trouble statistics and auto-open trouble tickets when user defined criteria are met.

Nonmanagement specific tools that can be integrated into the management infrastructure include e-mail, paging systems, and other notification systems. The management system will detect events and raise flags; the administrator can determine the notification procedures with these tools. As an example, an administrator may configure the system so that upon detection of a critical alarm an action request is sent to a paging system to immediately alert the shift supervisor. Systems may be configured so that upon detection of a threshold crossing, a complete PM history is generated and sent to a performance specialist via the e-mail system so that an investigation can be undertaken.

3.3.9. Application Programming Interfaces (APIs)

APIs are access points for software objects and are described by an Interface Definition Language (IDL). The IDL describes the objects and services that are supported by that object. When two software objects are linked together, it is through an API. APIs perform a specific network management function. As a rule, whatever a network operator can do from a GUI interface, an automated software program should be able to perform through an API. A single software process can have many APIs linking its functions to many other software processes.

The linkage that connects two independent processes is often called an *application association*. Associated processes can reside on a single computer or can be distributed across a networked computer complex.

APIs allow customers to develop software modules that extend the platform's functionality. APIs describe the parameters that are passed between two cooperating software processes and the functionality provided by the software. APIs do not describe how a specified functionality is achieved; thus, no intellectual rights are actually divulged. The API specification describes how a customer-written software program would call and link to the platform software. Such a specification explicitly details the parameters passed to the platform software, the requested action to be performed by the software, and the results returned. Often, such specifications include an example of how a customer would use the API.

Object-oriented techniques support the layering of the abstraction process, and each layered abstraction provides an API to the higher layer above responsible for the next level of abstraction. The more abstraction, the simpler to perform complicated management tasks. In the realm of platform technology, this concept is easy to exploit to the advantage of the network operator by layering platform services to reduce complexity without "reinventing the wheel." Each successive platform layer provides an IDL interface for the software making use of its services. Thus, developers building a management or agent application on top of a platform would only have to program the most abstracted object IDL and not the more complicated lower layer API.

APIs must interface correctly to use the functionality provided by the management platform. In an object-oriented system, this is not always easy to accomplish since the variety of network elements supported by a single management system may be quite diverse in technology and in functionality. When developing an application from scratch, a developer can build to the APIs of the platform system directly and can achieve a normalized level of functionality through a direct API interface. This is shown in Figure 3.7. If functionality extensions are desired to support specific functions in the application or if it is necessary to link two preexisting APIs together, an additional level of interface complexity is required. The two major technologies used to link and extend APIs are based on object adapter technology and manual extensions to autogenerated code segments. Object adapters are independent processes that sit between the application and the platform to provide a dynamic translation of object references between similar, but different, APIs. Such object adapters plug

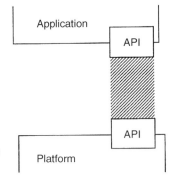

Figure 3.7 APIs Link Application to Platforms.

together objects via indirect references rather then forcing inherited relations to span across process boundaries. Code extensions are normally generated through the GDMO compile process. Many applications provide code stubs that link to user-specified object interfaces; developers manually tie the code stubs into the interface code before doing the final make/link application build. Object adapters are more fluid and permit interface independence between software modules, whereas code extensions can provide higher system performance.

X/Open Management Protocol (XMP) and X/Open OSI Abstract Data Manipulation (XOM) are the interfaces to the software layers that provide Association Control Service Element (ACSE) extensions. The XOM/XMP interfaces were defined by the X/Open Association. These are APIs that provide location-independent access to processes that are distributed through a OSI layered network. The XOM API is a computer industry standard API that supports the creation and manipulation of objects used by XMP. XMP is a general-purpose programming API that supports CMIS and SNMP service primitives. XOM defines the manipulated information objects as a C-type data structure. Manager applications must link to the XOM application using a defined data structure that is representative of the object being manipulated by the XMP request. XMP can be used to create, modify, or destroy the objects referenced by XOM. XOM allows objects to be referenced by object identifier or by object name. Upon completion, XMP will return a result status and an invocation identifier.

3.4. STANDARDS AND PLATFORM BUILDING BLOCKS

Network elements are evolving at an incredible rate. Certainly, no one would argue that innovative deployment of new network element technology should be slowed. Most of these evolutionary steps are driven by network operators who themselves are trying to differentiate among one another in a competitive environment. As network elements roll out new features to meet the needs of a dynamic market, the management system needs to keep pace with that evolution. Clearly, this is an impossible task for the network management supplier without the help of a sophisticated platform and set of support tools. The only possible solution to allow continued network evolution without precluding the ability to manage the network is to move to an environment in which the network element supplier has the ability to dynamically plug software modules into a platform environment that supports the latest achievements.

TMN standards provide models as generic guidelines and do not intend to inhibit network development by limiting each vendor to the features supported within those models. The intent is to define a relatively common set of what can be supported by all CMISE network elements and managed via a common mechanism. The intent is to encourage network element vendors to modify their object to add their individualized features. When vendors modify an object class, they are in effect defining a new class that needs to be registered publicly. Furthermore, since not all network elements will support all possible features, deletion of unsupported attributes will also drive new class definitions.

With a generalized platform, users can define their own network model with objects that provide customized service and operating characteristics for the carrier or use the modeling constructs supplied by the equipment vendor or by a generic modeling organization.

To meet these shifting operational demands, an independent network-level information model that provides a high-level abstraction of the network elements can be built over the object model that describes the physical characteristics of the network elements. In this fashion, a multiple object can be mapped into a single composite object. The same functions that operate on the physical objects can be applied to the composite logical objects to build a recursive relationship into their management structure. Such a recursive structural organization simplifies the network management problem significantly. Platforms provide a basis on which recursion may be built into the management infrastructure.

3.4.1. TMN

TMN defines several System Management Functional Areas (SMFAs), which together provide a complete network management framework. Fault management provides the ability to identify faults and supplies the diagnostic and test functions needed to isolate and validate problem-related activities. Configuration management creates, edits, and removes relationships between physically or logically related objects. Accounting management collects data that reflect who and when network resources are used. Performance management functions provide the ability to collect data that are used to statistically calculate network quality of service measurements. Security management functions are used to manage specific user access to data. It is important to realize that SMFAs are not applications but are functional groupings that can be supported within TMN applications. SMFAs provide a means of grouping functions into categories that simplify the description of a task. For instance, when users establish an end-to-end path through the network, they may activate new fault management points, change the configuration management database, activate accounting management collection, and establish performance management thresholds.

SMFAs can be divided into a series of functional components. As such, functional software modules that provide SMFA capabilities are candidates for inclusion in a platform architecture. Independent software modules in different SMFAs must work together to achieve a specific task. Since software products are often distributed on the basis of their ability to achieve a desired operation, most commercial software TMN systems include multiple modules that cut across SMFA boundaries. Such a structure will permit different operators to put together different software products and identify replicated functions and functional vacancies from a systems-level perspective. Defined APIs between SMFAs will allow partnered suppliers to work toward shared components, which then evolve from the application-specific category to a platform level. As this inevitable evolution begins to occur, operators will develop the ability to put the same software modules together in a different fashion to achieve operationally differing networks. Platform-based systems provide the modularized flexibility that is key in the migration to more competitive networks.

As an example, consider how the fault management SMFA can be subdivided into a series of platform modules and used to manage the network event notification process. Network events result in an object state change, including card, system, facility, and software failures. Since remote network elements may be managed from multiple sources (such as a local element operator), an alternate manager's modification of a provisioning parameter must result in a provisioning event being sent to the remaining managers in order to maintain consistency. Event management services, which are part of the fault management SMFA, include several functions such as event forwarding, event logging, event alerting, event trapping, and event display. The logging function provides a history of events that can be used to analyze network conditions. Event forwarding routes event notices to multiple manager processes which have registered as being interested in receiving those event indicators (and filters unwanted events from application with no interest). Event trapping monitors for specific defined network conditions and initiates predefined user actions when the requested conditions are met. The event alerting mechanism is used to draw operators to recent network actions that require manual intervention. Event display functions provide a means to allow the network operator to browse through current and historic network event reports.

Key management functions, all of which could be considered as candidates for inclusion in a platform architecture, might include the following:

Access Control Functions—support security controls on a per object basis across the MIB

Alarm Functions—support relating alarms to their causal classification

Audit Trail Functions—support the logging of audit trail data

Diagnostic Functions—support diagnostic routines that report operability and availability details

Event Report Functions—support the distribution of event reports

Knowledge Functions—support MIB structural information queries

Log Control Functions—support the collection and management of logging information

Metering Functions—provide support for logging usage data for accounting purposes

Monitoring Functions—provide a metric to manage object usage with thresholds for load balancing

Object Functions—support object creation, inquiries, modification, and deletion

Relationship Functions—support protected object relationships

Scheduling Functions—support the scheduling of periodic procedures

Security Alarm Functions—support the reporting of security breaches

State Functions—support modification of object state attributes

Summarization Functions—provide reports of overall status based on statistical data

Test Functions—support initiation and termination of test processes.

3.4.2. The NM Forum

The NM Forum is a consortium of manufacturers working together to close some loopholes in the existing management standards. The NM Forum has defined a series of Open Management Interoperability Points (ONMIPoints) that establish a baseline for open interworkings between software modules on a common or mixed-vendor platform. Omnipoint1 specifies management functions, including object, state, alarm and relationship management, scheduling, security, trouble, and path trace management functions. It also specifies 70 MOCs that should be supported from ISO standards, ITU standards, the NM Forum Library, and the OSI Implementors Workshop library. The goal of the OMNIPoint specifications is to achieve network-level information interworking when the information is assumed to be derived and supported by a management system and not to come directly from the network elements. It assumes that the network element support systems are object oriented and event driven.

The Service Providers Integrated Requirements for Information Technology (SPIRIT) is a team within the Network Management Forum. It was formed to provide for a collaborative international effort among telecommunications service providers to prepare, together with information technology (IT) suppliers and Independent Software Vendors (ISVs), a common set of software specifications for computing platforms. The aim of SPIRIT is to produce an agreed-upon set of specifications for a general-purpose computing platform for the telecommunications industry. A general-purpose computing platform, as shown in Figure 3.8, is a set of

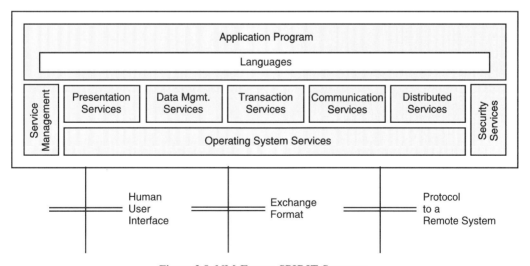

Figure 3.8 NM Forum SPIRIT Structure.

systems software specifications that is not aligned with a specific application type. The desired qualities of the general-purpose computing platform are portability, interoperability, and modularity.

The following classes of abstract machines are identified as the components of SPIRIT's view of a general-purpose platform:

- *Operating System Services*: The operating system services manage the fundamental physical and processing resources of a given machine.
- *Management Services*: The management services effect the changes of managed items on the platform.
- *Security Services*: The security services enforce security policies on data and processing objects on the platform and data objects exchanged between platforms.
- *Distributed Services*: These services consist of support services that facilitate cooperative processing between two platforms (for example, naming and distributed time services).
- *Communications Services*: Communications services emit and accept protocol. They may be layered on top of each other. Examples of communications services include TCP/IP, OSI Transport 0,2,4, and OSF DCE RPC.
- *Data Management Services*: These services manage persistent storage. They also presume some kind of data model. ISAM, CODASYL, and relational DBMSs are all examples of data management services.
- *Transaction Services*: Transaction manager services coordinate resources in order to maintain transactional integrity over those data resources. In order to do this, updates are applied in units of work called transactions, which are defined as having atomicity, durability, and serializability properties.
- *Presentation Services*: Presentation services act as the mediator between the system and human user interface devices (display, keyboard, mouse, etc.).

3.4.3. DCE

The Open Software Foundation (OSF) is working to develop guidelines that allow internetworking of heterogeneous open computer systems in a client-server environment. One of its projects focuses on developing an infrastructure to support distributed computer systems. The Distributed Computing Environment (DCE) project specifies an infrastructure platform that allows the software processes to be distributed over a wide area and still interoperate in a coordinated fashion via the API contracts. The DCE specification defines the services and tools needed to create, use, and administer a series of distributed applications in a distributed and open computer environment. A DCE-compliant platform allows a user to submit a request to one of many distributed server processes.

DCE concepts are based on the client-server application process relationship, depend on RPCs to provide the means for applications to communicate with each other, and have access to a common database organizational structure. A single distributed application is subdivided into a series of subprocesses that can be run on different computers. Servers are constructed to be independent of the location and number of client applications that may make use of their services. Server processes are normally implemented as an independent process (daemons).

DCE services provide a number of services (DCE threads) that are used to create, manage, and synchronize related application processes. Directory services allow the unique identification of resources within the distributed framework; time services synchronize activities on independent computers; security services provide access and authentication services for operators and for remote processes; distributed file services manage the file hierarchies spread across the interworking computer platforms; and other services define the properties of the local file system and the means to provide diskless support services.

In a scaleable network management system, an initial system can be deployed to manage a complete subnetwork of a given size and message volume. As that network grows, the subnetwork can be subdivided into multiple domains, and the processing load of the original system can be distributed across multiple processing platforms. Not only will the agent process be distributable in a TMN environment, but also the management processes should be distributable as well. This management process distribution is needed to allow the placement of manager processes to be determined by its own traffic loads and not by physical limitations driven by the placement of peer manager processes.

3.4.4. CORBA

The Object Management Group (OMG) Common Object Request Broker Architecture (CORBA) is specifically being developed to support the integration of object-oriented software in a heterogeneous distributed computer environment. Many of the goals of CORBA and DCE are very similar in nature. CORBA, however, is not targeted at a POSIX computer environment, and it assumes that a variety of different operating systems will be interworking to accomplish the requested task. The CORBA paradigm assumes an object-oriented environment, with object location transparency being a prime requirement. This drives the requirement for the Object Request Broker (ORB), which is a key differentiator within the CORBA environment.

Like DCE, CORBA provides the facilities needed to support a scaleable distributed computing infrastructure. Objects which reference stored or dynamically managed data can be viewed as persistent, or they can be software processes that can act on a user's behalf. Such processes can be managed across the distributed infrastructure boundaries as in a DCE environment. While DCE is a more mature concept than CORBA, many observers expect that CORBA will eventually eclipse DCE in functionality and deployment.

3.5. CASE STUDY: OPENVIEW, ALMAP, AND THE 1320

Many examples of commercial tools, platforms, and middleware are available for integration into a TMN platform structure. Generally, the larger the networking requirement and the more diverse the support requirements, the more complicated and the more important the management platform becomes. In an effort to illustrate the concepts described, it may be of value to discuss one such complicated implementation scenario.

Alcatel is a large supplier to the communications industry with products that cover a wide range of technologies and application environments. It has product development efforts spread across the globe, working on products that meet a wide variety of network operator requirements. Since all these intelligent network elements need some form of management, Alcatel is rapidly moving in the direction of platform-based technologies to harness and reuse as much of this energy as possible. Before TMN, a multifunctional management application was nearly impossible, and as the management standards continue to evolve and become realizable, Alcatel is using platform-based technologies to increase the leverage of management efforts.

The Alcatel effort began with an overriding network management architecture called ALMAP. ALMAP builds on top of HP OpenView as the basic platform to achieve its network operator TMN environment. Alcatel began by structuring two ALMAP compliant metalayers within the Alcatel platform structure. One ALMAP metalayer provides application modules specifically targeted for use within Alcatel for the development of SDH network applications. The 1320 NM metalayer is a commercialized product that provides middleware and application platform modules which a network operator or a systems integrator can use to extend the functionality of the management system to meet individualized requirements.

3.5.1. HP OpenView

HP OpenView is a family of integrated network and system management solutions for managing the complete information technology enterprises, including networks, distributed systems, and applications. The HP OpenView DM Platform furnishes a common graphical user interface and a standards-based communication infrastructure that underlies all telecommunications management applications.

OpenView is built on the Communications Infrastructure (CI), protocol stacks, and related components: GDMO Meta-data, Event Management Services, and Object Registration Services. The DM Platform includes the same HP OpenView Windows (OVw) user interface supplied with the HP OpenView SNMP platform, Network Node Manager, and Operations Center products.

The HP OpenView Distributed Management (DM) Platform that underlies the application components consists of a number of main building blocks:

- The Communications Infrastructure serves as the integration point for management protocol stacks such as CMIP and SNMP, management APIs, and related facilities stacks (routing, events, and association control). The Communications Infrastructure provides distributed message routing and access to applications and services through standard management protocols.

- Event Services collect, forward, and log network events and traps in a distributed environment. This module provides event routing, filtering, logging, and administration. It is also extended to include SNMP.

- Object Services and Tools provide a GDMO (Guidelines for the Definition of Managed Objects) tool chain that compiles, stores, and provides dynamic access of object definitions. Meta-data (the definition or structure of objects) retrieval allows applications to dynamically interact with new objects.

- HP OpenView Windows is a Graphical User Interface (GUI) that provides network operators and administrators with a consistent view of the managed environment and seamless integration of management functions regardless of vendor or managed object type. HP OpenView Windows provides a common interface that simplifies both the development and use of management applications.

The HP OpenView Distributed Management Developer's Kit provides all the Application Programming Interfaces (APIs) for interacting with HP OpenView Windows and management protocols via XMP (X/Open Management Protocol) and XOM (X/Open Object Management) APIs. The XMP API provides programmatic access to both SNMP and CMIS services, the Event Management Services, and meta-data. XMP also includes both a GDMO-based object definition package generator to simplify application development and programmatic control over the association control services element (ACSE). An XOM function generator and a hypertext on-line reference system are included in the Developer's Kit along with an ASN.1 compiler. Process control, tracing, and logging APIs are included so that applications can consistently integrate with HP OpenView's process control and troubleshooting procedures. Utility routines and sample applications are included for developers. A lightweight API is supplied for creating SNMP-only management applications. The XMP and XOM APIs can be used for building portable, standards-based manager, agent, and combined manager/agent applications.

The HP OpenView GDMO Modeling Toolset is used to analyze and design OSI object models. The tools allow administrators to graphically design and edit object specifications and consists of the following building blocks:

- The GDMO editor/browser is the primary tool for OSI NM object modelers. It allows operators to create, edit, and browse GDMO definitions interactively through a graphical user interface.

- The Object Dictionary is the central repository for the managed object (MO) classes created with the editor/browser and the import tool. It supports concurrent access to data from multiple operators. It also contains built-in libraries of OSI-standard MO classes, such as X.721, Q.821, and M.3100.

- The GDMO Import Tool takes an ASCII input file containing the GDMO and ASN.1. definitions, performs intratemplate syntax and semantic checking, and stores the definitions in the object dictionary.
- The GDMO Export Tool produces an ASCII file containing all the GDMO templates and ASN.1 definitions belonging to a given document.
- The Document Manager maintains GDMO documents within the object dictionary and allows users to partition GDMO definitions into one or many logical documents.
- The GDMO Semantic Checker validates an entire document by assuring that all GDMO rules are adhered to.
- The Conformance Report Generator automatically produces conformance statements consistent with the ISO/IEC IS 10165-6 standard.

The Managed Object Toolkit (MOT) automates a portion of TMN application development, accelerating the programming task by providing an agent application infrastructure and a generated C++ agent code skeleton on which applications can be built. The MOT infrastructure also handles the transmission and receipt of CMIS transactions. The C++ class libraries supplied with the MOT provide an OSI standard conformant implementation of the CMIS services (for example, get, replace, scoping, filtering, linked replies, and service errors).

The Managed Object Toolkit (MOT) consists of two major parts:

- A C++ Code Generator, which provides DM application programmers with C++ classes that the developer can utilize to develop agent or manager applications. One subset of the MOT-provided classes encapsulates XOM object manipulations for manager developers when preparing CMIS requests.
- The Agent Infrastructure, which handles CMIS request and prepares CMIS responses for the agent application, insulating the agent developer from the XOM/XMP (X/Open Management Protocols) API and the CMIS service model.

Distributed communications are managed through the HP OpenView Postmaster. Any manager or agent application may register with the postmaster to receive any notifications received by the TMN system. The postmaster keeps track of all registered event forwarding discriminators (EFD) for each such application. Multiple manager/agent applications may register to receive a single message, and by not registering, managers can use the postmaster to filter events from their queues. To register, the registering application has to give the postmaster a discriminator construct and a destination address. Once accepted, any event notifications that match the template established by the discriminator construct are forwarded to the destination process.

Event Management Services (EMS) models the ISO definitions for alarms and logs, GDMO-based meta-data services for compiling and on-line access of object definitions, and the Object Registration Services (ORS) which provide distributed knowledge of managed objects, instances, and physical addresses. By using ORS, an Application developer does not need to know where a particular Managed Object is located or to which Agent it belongs. ORS takes care of routing requests to the MO, wherever it may be in the network (on the same server or on a remote one). EMS enables managers to subscribe to specific events through creation of event forwarding discriminators, without having to know which agent can send them.

The OpenView DM Platform is used as a subplatform for the HP OpenView SNMP Node Manager and a range of other OpenView applications, such as the Operations Center. The HP OpenView Distributed Management (DM) products supply a standards-based environment for building a network management system. It provides multiple protocols (CMIP and SNMP) for building OSI and mixed protocol solutions. HP OpenView is built on the open systems concept, enabling it to run on a variety of hardware platforms.

HP OpenView Windows (OVw), a graphical map manager, provides the map integration of Applications, that is, the capability to make selections on the map representing network entities, and to navigate from these icons to specific applications either through double-clicking or using tailored pull-down menus.

3.5.2. ALMAP

Alcatel's common network management platform, called ALMAP, builds on top of the functionality provided by HP OpenView. The ALMAP platform defines a number of Generic Components (GCs), each of which provides a specific type of platform services. Although the initial releases of the GCs were targeted specifically at the support of the SDH management environment, they are being extended to increase their functionality in order to become more general purpose. The GCs include the following:

The Distributed System Manager (DSM) GC provides process, application, and system management functionalities. Applications in the sense of DSM consist of a set of interacting, distributed processes. These processes are managed as a single entity. Supervision, start/stop, and restart of applications observe the dependencies of processes and their temporal order. The core component of DSM, called the universal process manager (UPM), is instantiated on each computer node. It contains several specific submanagers. Thus, DSM is able to manage different process types. Managers for DCE processes, Daemons, HP OV processes, and general Unix processes are available. The Base Component is open to integrate further managers to handle other specific process types, for example, for the management of legacy systems.

Security Management (SEC) GC aims at providing generic mechanisms for managing security information. SEC covers six relevant sections:

- User Identification and Authentication (based on Unix mechanisms).
- User Session Control (based on Unix mechanisms).

- Security Alarm Registration and Audit logs.
- Administration of user access control information.
- Distribution of user access control information across a distributed computer system.
- Via an API an application's request for a decision for access on a target object or function.

The Distributed Network Map Manager (DNM) GC permits the sharing of OVw maps between different operators, on a distributed network. DNM offers the following services:

- Store Map Image: The administrator stores a map and makes it visible to all or some operators. An e-mail message is sent to each of them for notification.
- Remove Map Image: The administrator can select from a list the map(s) to be removed.
- Modify Username List: The administrator can change the list of operators who can access a map.
- List/Get Map Image: The user is provided with the list of maps he or she is allowed to access, and can select one of them. The DNM then retrieves the map image from MAP DB and populates the OVw map.

The Alarm Surveillance GC is a management service common to many management domains of the TMN. Its main objective is the early detection of faults before significant effects have been felt by the end-user of the telecommunication system, detecting malfunctions, identifying faulty network elements, and providing information that will help with understanding the cause of potentially abnormal situations and will support the management operator on diagnostics and associated corrective actions. The Alarm Management GC supports alarm surveillance services and provides different capabilities needed to monitor or interrogate network elements (or both) about alarm events or conditions, providing that all the network element agents intended to use Alarm Management services, support a common management interface. The main functions covered are

- Alarm reporting management: it supports reporting and reporting control (sorting and filtering) of alarms and associated information.
- Internal storage and processing of current and historical alarm information, including alarm status processing (acknowledge, manual clear, transfer to archival log files).
- Alarm Event logging and log records retrieval and associated display services.

The Event and Log Management (ELM) GC comprises a process that is functional to a TMN manager Application dealing with instances of managed object event forwarding discriminator (EFD) and log. This component primarily offers a graphical user interface to manipulate EFDs and logs and maps the performed user actions to appropriate CMIS requests. ELM provides the following features:

- Configuring complex filtering constructs for distributing and logging alarm and event notifications.
- Presentation of lists of sorted and filtered log records as well as detailed log record information.

The Physical Network Management (PNM) GC is an equipment management layer component that provides two main functions: the integration of EM subsystems and the management of a network element directory. PNM offers the following services:

- Management of physical network maps: management of OVw maps, whose submaps group network elements from a geographical or topological perspective.
- Management of network element supervision: OVw submaps are animated according to network element states, and the operator can manage the processes that start, stop, and reset supervision processes.

The TMN OS Configuration Management (TCM) GC provides a controlled storage of configuration data for different communication processes as well as appropriate data usage.

3.5.3. Alcatel 1320 NM

The 1320 NM is a platform metalayer that, in the true sense of open systems, can be used with or without the ALMAP GCs. The 1320 NM provides a vendor-independent TMN network/element management platform. It was developed by Alcatel as an open platform for multivendor TL1 network element management, and it includes tools that simplify the development of the applications. These processes can be used to extend the functionality of the management system and to provide functionality that is particular to that network operator, or they can integrate support software for network elements from other equipment suppliers. The structure of the 1320 NM allows developers to incorporate these functional extensions into the 1320 NM platform umbrella in such a manner that the presentation to the network operator's operational personnel appears as seamless.

For example, the Alcatel 1320 NM builds on top of the HP OpenView platform to customize its services for a TL1 and CMISE network operator environment. The Alcatel 1320 Open Framework links from the HP Openview XOM/XMP API to present a CMISE API to the manager applications that build on it.

As part of the Alcatel strategy to provide a multivendor bridging system that spans legacy TL1 network elements into the TMN environment and on into a TINA environment, the 1320 NM includes support for composite logical objects. In the 1320 NM envisioned environment, the network management layer will generate high-level commands that act on logical composite objects and pass them to the appropriate EML. The EML allows local operators to manipulate the network via a GUI and will also allow the NML to manipulate the network. In response to operator request or a NML request, the EML will create the detailed messages needed to execute those commands. The EML will use provisioning commands to allocate time slots, assign features, activate cross connects, install and test customer services, and manage access to the data stored in network elements (e.g., performance). The EML will monitor and correlate alarms generated by network elements within a node group and send up to the Network Management Layer only those alarms that are needed to trigger maintenance activity. The NML manages the EML as a logical composite object that is capable of very intelligent resource management. The EML manages the mapping from the logical composite object to the physical resources that are within its span of control. This logical abstraction process frees the NML from having to be aware of the physical composition of the subnetwork processes and allows it to manage the network as a logical entity.

The 1320 includes a GDMO compiler that is specifically designed to support TL1 network elements. It is used to compile a tabular description of a network element. The compiler produces a GDMO description of the network element, loads the MIB browser, and generates the complete GUI drivers for a vendor-independent management interface and a skeleton C++ data structure that characterizes the network element object classes. These descriptive data structures are then linked with the 1320 proxy agent communications management software, the generic object management software, and TL1-CMISE conversion scripts to generate a fully functional proxy agent. The proxy agents act as object servers for the object-oriented management applications. As such, they manage all application requests and responses for network element directed activities, as well as the instantiation and object deletion processes, and oversee the object relationship linkages with nonelement-based objects.

As a CMISE-based platform built on a foundation of TL1-CMISE proxies, the 1320 NM will support the graceful migration of TL1 network elements to a CMISE environment. Network element managers can manipulate CMISE- and TL1-based network elements in an equivalent fashion; in the case of a CMISE network element, the proxy agent process is simply eliminated from the data flow. The 1320 NM was also designed to support numerous network elements of a similar and dissimilar nature. That is, the 1320 NM will support numerous lightwave terminal types from many vendors. In addition, the 1320 NM will support multiple network element types, including radio, cross connects, and STM/ATM multiplexers. The user will manage all supported network elements in a similar fashion. This is a critical concern because over time network element technologies will become blurred. Cross connects with optic terminations are already in the field, and lightwave devices with ATM support will soon be a reality. A technology-specific management system will be limited by the life of that technology and cannot evolve over time.

All the ALMAP components, including the 1320 NM, use a hardware and software structure based on a client-server structure in order to ensure easy extendibility. This allows the system to grow beyond the performance limitations of a single computer as the needs of the network operator grow.

The 1320 NM provides a working OSI Q3 protocol stack with CMISE to allow higher order management systems to have full-access TL1 network elements. It also provides a working CORBA interface to the intelligent composite objects, which allows the higher order systems an additional degree of power and latitude in managing the network elements within a 1320 NMs charge. An alarm management system, a MIB browser, a TL1, and a CMIS terminal are also provided as common application functions.

3.6. SUMMARY AND OUTLOOK

When network operators began to centralize operation support centers in the 1970s, many proprietary systems were introduced which attempted to automate existing operational practices. Such early systems were slow to roll out to production and difficult to maintain. Changes in operational procedures required lengthy development cycles. Support of new or upgraded network elements was time consuming, especially if they supported an unanticipated technical advance.

TMN concepts evolved to open up the internal structure of the network management system. The TMN architecture that underlies most current management platforms started slowly as a theoretical concept and has progressed rapidly as network operators attempt to deploy early systems.

Many early TMN platforms provided limited content. These early offerings were limited to a few modularized components that are easily defined and could be readily bundled into a defined product structure. These systems include:

Communication Services: Computer systems and network elements based on different technologies have to be interconnected. The communications services platform provides the communications protocols needed to allow the systems to exchange messages.

TMN Object Handling Services: TMN technology is strongly object oriented, and a TMN Object Handling package became a natural requirement of the TMN platform.

Infrastructure Kernel Services: Network equipment has to be scaleable in a wide scope. It also has to fulfill strong availability/robustness requirements. The Infrastructure Kernel Services supports process distribution in a managed client server arrangement.

The complexity of current network management systems is exponentially related to the operator process complexities, the number of network elements, the variety of network elements, and the rate at which these factors evolve over time. As network elements become more software-based and as competition continues to increase the rate of service evolution, we find the problems faced by network management systems increasing at a staggering rate. No single solution exists which will provide an answer for a network carrier looking to keep control of their network. A strong platform-based orientation is the only solution available to overcome the enormous difficulties facing those of us tasked with integrating a management system into a production environment.

The main role of the management platform is to reduce development costs and to provide application consistency. As the network elements become more intelligent, the software processes that are used to manage them will also become more intelligent. Even more so, the management applications will intelligently abstract such data and provide value-added functionality in addition. This evolutionary trend will continue until the entire network can be viewed as a series of object-oriented applets that functionally build on each other.

Not all of these evolving management functions will be standardized as carriers search for mechanisms to differentiate themselves in an increasingly competitive environment. For network operators and service providers, use of a platform with published APIs permits the addition of applications produced by them (or a third party). These system-specific extensions allow operators to distinguish their services by the nature of their management suppliers and by the incorporation of customized functionality into the web of interacting software products.

Commercial TMN platform providers must balance between conflicting evolutionary drivers. If platform vendors try to add application-level functionalities as much as possible, they increase the specificity of TMN product development significantly and decrease their product market. If platform vendors try to limit their work areas common across a mass market, the value of a platform in terms of saved effort is reduced.

TMN technology is very complex. Thus, the tool chains have a strong influence on the productivity of development teams. The current early-generation platforms concentrated on providing the ability to populate a standard meta-data agent. Although these tool chains did much to improve productivity, much more work can be done in this area. The next generation of tools which are just beginning to appear in the market promise to improve the amount of automation available and to improve the flexibility needed to support a diverse network element population. Higher layer tools will improve ability to rapidly generate manager applications and to hide the complexity of the low-level APIs. These emerging platform systems will increase the programming productivity significantly and will aid in the distribution of management intelligence further out into the network infrastructure.

Mobile networks and dynamic network infrastructures are beginning to form and will become the basis for a substantial amount of future-oriented network management direction. To accommodate the needs of the evolving network much attention will be given to the support of a more dynamic network infrastructure. This includes support for transportable and transient network objects that move between supporting object servers according to the network need. Web-based viewers and object broker technology will increase in importance to facilitate support of such active object structures.

All major platform vendors intend to move toward providing Internet-Intranet and CORBA compliant versions of their platform products starting within the 1997–1998 time frame. CORBA is an object-oriented, machine to machine communications mechanism developed to support interprocess communications. It promises to provide a more preformant and more cost-effective communications vehicle to link

TMN computer systems together. Object-oriented Web viewer systems such as Java produce a computer-independent mechanism for distributed operators to examine the permissible portion of their network. Complementing this entire trend is the rapidly evolving world of the object-oriented database. Such systems are rapidly reaching a maturity level where they can be counted on to support object-structured and scaleable environments demands by the telecommunications management industry. Since CORBA, Java, and object-oriented databases are all especially well suited for support of an object-oriented manipulation system, the convergence of these technologies will drive an evolutionary step that can be considered another jump in technology.

References

[1] Alcatel Network Systems, *1320 Technical Description*, internal draft release, 1995.

[2] Foody, Michael, "Making Objects Deliver in the Client/Server World," *Object Magazine*, March 1996, pp. 63–74.

[3] Guadagno, Luigi, and Okon, Okokon, "Stepping Up the Flow of Business Information," *Object Magazine*, July 1996, pp. 41–45.

[4] Hegering, Heinz-Gerd, and Abeck, Sebastian, *Integrated Network and System Management*, Addison-Wesley Publishing, Co., 1994.

[5] Hewlett Packard, *HP OpenView Platform Technical Evaluation Guide*, 1994, 5962-9587E.

[6] Pree, Wolfgang, "Frameworks—Past, Present, Future," *Object Magazine*, May 1996, pp. 24–26.

Vijay K. Garg*
Lucent Technologies,
Bell Labs

Chapter 4

Management of Personal Communications Services (PCS) Networks

4.1. INTRODUCTION

The goal of wireless communication is to allow the user access to capabilities that are presently available with wireline networks at any time, anywhere, regardless of the access method. Cellular and cordless telephony [1], both of which have gained widespread user acceptance during the last decade, started this process, but do not yet permit total wireless communications. Personal Communications Services (PCS) is an attempt in that direction. We can regard cellular and cordless systems as the earliest version of wireless PCS. Cellular systems allow the subscriber to place and receive telephone calls over the wireline telephone network wherever cellular coverage is provided. "Roaming" capabilities allow services to users traveling outside their "home" service area.

Early analog cellular systems are being replaced by digital systems because digital systems can support more users per base station per MHz of spectrum. Two digital technologies, Time-Division Multiple Access (TDMA) and Code-Division Multiple Access (CDMA), are emerging as clear choices for a PCS system. TDMA is a narrowband system in which communications per frequency channel are apportioned according to time. For TDMA systems, there are two prevalent technologies: (1) North American TIA/EIA/IS-136 [2] and (2) European Telecommunications Standards Institute (ETSI) Digital Cellular System 1800 (DCS1800) [3] similar to Global System for Mobile Communications (GSM). CDMA (TIA/EIA/IS-95) [4] is a direct sequence spread spectrum system in which

*Presently working at Motorola Inc., Arlington Heights, IL.

the entire bandwidth of the system is made available to each user. The bandwidth is many times larger than the bandwidth needed to transmit information. CDMA systems are limited by interference. Pseudorandom sequences for the different user signals, with the same transmission bandwidth, are used in CDMA systems.

In a PCS network, users and terminals are mobile rather than fixed as in wireline networks. A PCS network is designed to provide personal, terminal, and service mobility. *Personal mobility* is the ability of users and telecommunications services to remain mutually accessible regardless of location. A user can access telecommunications services at any terminal on the basis of a personal identifier, and the network can provide those services according to the user's profile. A user can access services from a wireline or wireless terminal. *Terminal mobility* allows a terminal to access telecommunications services from different locations while in motion. *Service mobility* deals with the ability to use vertical features from remote locations or while in motion. Although terminal mobility has been mostly emphasized for PCS, all three elements of mobility will be required to meet the increasing needs of users.

In the past, management of telecommunications networks was well understood and straightforward. Before deregulation and privatization of the telephone industry, there were fewer issues to deal with, and generally the competitive pressure was less intense. Telecommunications networks encompassed equipment from fewer vendors, and so there were fewer multivendor management issues. As service providers moved into a mixed-vendor environment, they could no longer afford different management systems for each network element (NE). To address this issue, standards groups in Committee T1 and ITU-T defined the interface and protocols for operations support systems under the umbrella of Telecommunication Management Network (TMN) (see Chapter 1 for details of TMN) [5, 6, 7].

Service providers are deploying PCS into an environment where there will be intense competition with existing cellular companies. Therefore, they must offer cost-effective services. While management functions are necessary for the smooth operation of a network, they are also a cost of doing business. However, a superior Operation, Administration, Maintenance, and Planning (OAM&P) should improve the bottom line of the business by reducing network operational costs. Service providers must manage the cost of OAM&P. Considerations of initial cost of management systems and their operational cost must be carefully examined. A key to success of PCS networks lies in the system operator's ability to effectively manage new equipment and services in mixed vendor environments. For PCS to succeed, the OAM&P costs of PCS networks and services must be comparable or less than those of either the existing wireless or wireline networks.

Automation is essential to develop machine-to-machine interfaces that will replace many manual functions. The need to manage the equipment of mixed vendors requires some form of standardization. The need to support rapid technological revolution suggests that interfaces to the management systems be general and flexible [8, 9, 10]. To ensure that the interfaces have sufficient consistency to allow some level of integrated management, it is necessary to develop a set of guidelines. Two approaches to manage a PCS network have been suggested: one based on Simple Network Management Protocol (SNMP) and the other based on Open System Interconnect (OSI) Systems Management.

In this chapter, we address the requirements of those OAM&P elements of a PCS network that keep the system operating on a day-to-day basis. With OAM&P systems, PCS service providers monitor the health of all NEs, add and remove equipment, test software and hardware, diagnose problems, and bill subscribers for services.

The following section presents management approaches for the PCS network and discusses SNMP- and CMIP-based management schemes. In Section 4.3 we include two North American reference models for PCS networks and discuss the interfaces between NEs. The requirements for PCS network management and management goals are given in sections 4.4 and 4.5. In section 4.6, we outline the high-level functional requirements in the five functional areas of management.

4.2. MANAGEMENT APPROACHES FOR THE PCS NETWORK

SNMP [11] is a basic, easily implementable network management tool for the Transmission-Control Protocol (TCP)/Internet Protocol (IP) environment. The SNMP framework includes the structure of management information, which decides the format of management information, as well as Management Information Base (MIB). MIB is the current standard database for SNMP management information. Within the framework of SNMP, additional MIBs can easily be defined to support OAM&P functions of a PCS network. A major enhancement to SNMP, known as SNMP-2, has recently been standardized.

Open System Interconnect (OSI) Systems Management has been derived from the set of guiding principles provided by TMN. OSI Systems Management uses Common Management Information Service (CMIS) for a basic network management service to network management applications. The Common Management Information Protocol (CMIP) is used to implement CMIS in the OSI Systems Management (refer to Chapter 1 for CMIS and CMIP).

TMN includes a logical structure that originates from data communication networks to provide an organized architecture for achieving interconnection between various types of operations systems (OSs) or telecommunications equipment types. Management information is exchanged between OSs and equipment using an agreed-upon architecture and standardized interfaces, including protocols and messages. The principles of TMN provide for management through the definition of a management information model (Managed Objects) [12] which is operated over standardized interfaces. This model was developed through the definition of a required set of management services which are divided into various service components and management functions. An information model represents the system and supports the management functions. Three important aspects of TMN which have been applied to manage PCS network are:

- A layered architecture with five layers: Business Management Layer (BML), Service Management Layer (SML), Network Management Layer (NML), Element Management Layer (EML), and Network Element Layer (NEL). A layered architecture, as applied to a PCS network, is shown in Figure 4.1 (refer to Chapter 1 for details).

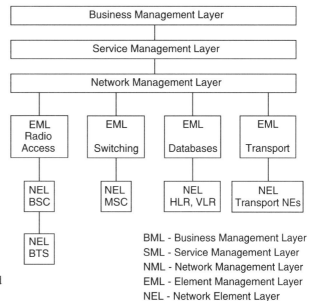

Figure 4.1 Layered Architecture Applied
                    to PCS Network.

BML - Business Management Layer
SML - Service Management Layer
NML - Network Management Layer
EML - Element Management Layer
NEL - Network Element Layer

- A functional architecture (see Figure 4.2) that defines functional blocks: Operations Systems Function (OSF), Mediation Function (MF), Work Station Function (WSF), Network Element Function (NEF) and Q Adapter Function (QAF).

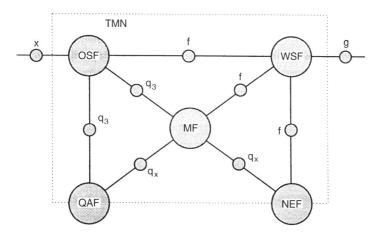

WSF = Workstation Function
OSF = Operations System Function
MF   = Mediation Function
NEF = Network Element Function
QAF = Q Adapter Function

Figure 4.2 TMN Functional Architecture and Interfaces.

- A physical architecture defining management roles for OSs, communication networks, and network elements (NEs).

Table 4.1 summarizes the major differences between SNMP-2 and CMIP.

Comparison of SNMP-2 and CMIP in Table 4.1 shows that CMIP is superior and more flexible as compared to SNMP. Both TIA and ETSI have selected CMIP for management of PCS and Future Public Land Mobile Telephone System (FPLMTS). For greater details, see Chapter 2.

TABLE 4.1   SUMMARY OF MAJOR DIFFERENCES BETWEEN SNMP-2 AND CMIP

Comparison criteria	SNMP-2	CMIP
Object-oriented	Not object-oriented, change on objects may impact applications.	Objects encapsulated, change on objects does not impact applications
Agent/manager	Manager-to-manager relationship possible (informrequest)	Based on the TMN architecture
Scoping	Bulk retrieval possible without using get next	Can define subtrees; several agents may have the same address
Counter size	64 bits	64 bits
Functions	Trap, get request, get-next request, set request, get response, get bulk, informrequest	M-event-report, M-get, M-cancel-get, M-action, M-create, M-delete, M-set
Security	Secure SNMP modified to improve performance	Encryption, user validation, and OSI protocol features
Transport protocol	Connectionless	Connection-oriented
Events collections	Polling or bulk polling to get traps	Event driven, a state change generate notification
Flow control	No source control flow	Source control flow
Alert format	Six different types of traps, sixth category is enterprise specific	Generic format for alerts

4.3. REFERENCE MODEL FOR THE NORTH AMERICAN PCS SYSTEM

Key to the North American PCS System is the use of a common reference model. Both the TIA (in TR-45 and TR-46) and Committee T1 (T1P1) have a reference model, and each model can be converted into the other. All of the North American PCS systems follow the reference model. The names of each NE are similar, and some of the functionality is partitioned differently between the models. The main difference between the two reference models is how mobility is managed. In the T1 reference model, user data and terminal data are separate; thus, users can communicate with the network via different Radio Personal Terminals (RPTs). In the TIA reference model, only terminal mobility is supported. A user can place or receive calls at only one terminal (the one the network has identified as assigned to the user). TIA functionality is migrating toward independent terminal and user mobility, but not all aspects of it are currently supported. Although these models are for PCS, they apply equally well to cellular systems.

4.3.1. TIA Reference Model

The main elements of the TIA reference model [13] (see Figure 4.3), are

- **Mobile Station (MS):** The MS terminates the radio path on the user side and enables the user to gain access to services from the network. The MS can be a stand-alone device, or it can have other devices (e.g., personal computers, fax machines, etc.) connected to it.

- **Radio System (RS):** The RS, often called the Base Station, terminates the radio path and connects to the Personal Communications Switching Center. The RS is often segmented into the BTS and the BSC:

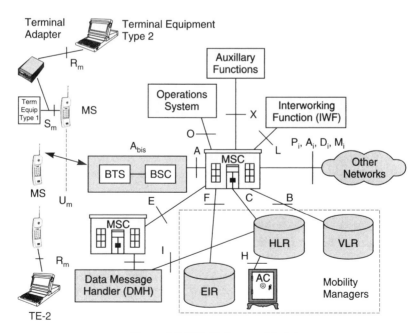

Figure 4.3 TIA Reference Model.

- **Base Transceiver System (BTS):** The BTS consists of one or more transceivers placed at a single location and terminates the radio path on the network side. The BTS may be co-located with a BSC, or it may be independently located.

- **Base Station Controller (BSC):** The BSC is the control and management system for one or more BTSs. The BSC exchanges messages with both the BTS and the MSC. Some signaling messages may pass through the BSC transparently.

- **Mobile Switching Center (MSC):** The MSC is an automatic system that interfaces user traffic from the wireless network to the wireline network or other wireless networks. The MSC is also often called the Mobile Telephone Switching Office (MTSO) or the Personal Communications Switching Center (PCSC).

- **Home Location Register (HLR):** The HLR is the functional unit used for management of mobile subscribers by maintaining all subscriber data (e.g., ESN, DN, IMSI, user profiles, current location, etc.). The HLR may be co-located with an MSC, an integral part of the MSC, or it may be independent of the MSC. One HLR can serve multiple MSCs, or an HLR may be distributed over multiple locations.

- **Data Message Handler (DMH):** The DMH is used for generating data employed in billing the customer.

- **Visited Location Register (VLR):** The VLR is linked to one or more MSCs. The VLR is the functional unit that dynamically stores subscriber information, (e.g., Electronic Serial Number (ESN), Directory Number (DN), User Profile Information) obtained from the user's HLR, when the subscriber is located in the area covered by the VLR. When a roaming MS enters a new service area covered by an MSC, the MSC informs the associated VLR about the MS by querying the HLR after the MS goes through a registration procedure.

- **Authentication Center (AC):** The AC manages the authentication or encryption information associated with individual subscribers. The AC may be located within an HLR or MSC, or it may be located independently of both.

- **Equipment Identity Register (EIR):** The EIR provides information about the mobile station for record purposes. The EIR may be located within an MSC, or it may be located independently of it.

- **Operations System (OS):** The OS is responsible for overall management of the wireless network.

- **Interworking Function (IWF):** The IWF enables the MSC to communicate with other networks when protocol conversions are necessary.

- **External Networks:** These are other communications networks: the Public Switched Telephone Network (PSTN), the Integrated Services Digital Network (ISDN), the Public Land Mobile Network (PLMN), and the Public Switched Packet Data Network (PSPDN).

The following interfaces are defined between various elements of the system:

- **RS to MSC (A-Interface):** The interface between the Radio System and the MSC supports signaling and traffic (both voice and data). A-interface protocols have been defined using SS7, ISDN BRI/PRI, and frame relay.
- **The BTS to BSC Interface (A-bis):** If the Radio System is segmented into a BTS and BSC, this internal interface is defined.
- **The MSC to PSTN Interface (A_i):** This interface is defined as an analog interface using either Dual Tone Multifrequency (DTMF) signaling or Multifrequency (MF) signaling.
- **MSC to VLR (B-Interface):** This interface is defined in the TIA IS-41 protocol specification.
- **MSC to HLR (C-Interface):** This interface is defined in the TIA IS-41 protocol specification.
- **HLR to VLR (D-Interface):** This interface is the signaling interface between an HLR and a VLR and is based on SS7. It is currently defined in the TIA IS-41 protocol specification.
- **MSC to ISDN (D_i-Interface):** This is the digital interface to the Public Telephone network and is a T1 interface (24 channels of 64 kbps) and uses Q.931 signaling.
- **MSC-MSC (E-Interface):** This interface is the traffic and signaling interface between wireless networks. It is currently defined in the TIA IS-41 protocol specification.
- **MSC to EIR (F-Interface):** Since the EIR is not yet defined, the protocol for this interface is not defined.
- **VLR to VLR (G-Interface):** When communication between VLRs is needed, this interface is used. It is defined by TIA IS-41.
- **HLR to Authentication Center (H-Interface):** The protocol for this interface is not defined.
- **DMH to MSC (I-Interface):** This interface is defined by IS-124.
- **MSC to the IWF (L-Interface):** This interface is defined by the interworking function.
- **MSC to PLMN (M_i-Interface):** This interface is to another wireless network.
- **MSC to OS (O-Interface):** This is the interface to the operations systems. It is currently being defined in ATSI standard body T1M1.
- **MSC to PSPDN (P_i-Interface):** This interface is defined by the packet network that is connected to the MSC.
- **Terminal Adapter (TA) to Terminal Equipment (TE) (R-Interface):** These interfaces will be specific for each type of terminal that will be connected to an MS.
- **ISDN to TE (S-Interface):** This interface is outside the scope of PCS and is defined within the ISDN system.
- **RS-PS (U_m-interface):** This is the air interface.

- **PSTN to DCE (W-Interface):** This interface is outside the scope of PCS and is defined within the PSTN system.
- **MSC to AUX. (X-Interface):** This interface depends on the auxiliary equipment connected to the MSC.

4.3.2. T1 PCS Reference Architecture

The T1 Architecture [14] (Figure 4.4) is similar to the TIA model but has some differences. The following elements are defined in the model:

- **Radio Personal Terminal (RPT):** The RPT is identical to the MS of the TIA model.
- **Radio Port (RP):** The RP is identical to the BTS of the TIA model.
- **Radio Port Intermediary (RPI):** The RPI provides an interface between one or more RPs and the RPC. The RPI allocates radio channels and may control handoffs. It is dependent on the air interface.

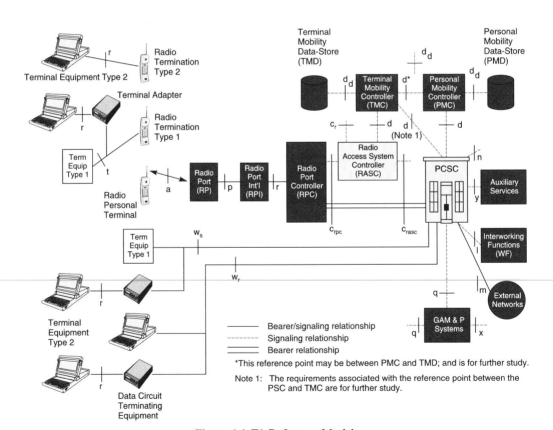

Figure 4.4 T1 Reference Model.

- **Radio Port Controller (RPC):** The RPC is identical to the BSC of the TIA model.

- **Radio Access System Controller (RASC):** The RASC performs the radio-specific switching functions of Call Delivery and Origination, Handoff control, Registration and Authentication, and Radio Access Management (control of signaling channels).

- **PCS Switching Center (PSC):** The PSC is similar to the MSC of the TIA reference model. Some of the functions of the MSC in the TIA model are distributed into other elements.

- **Terminal Mobility Controller (TMC):** The TMC provides control logic for terminal authentication, location management, alerting, and routing to RPTs. This function is supported by the VLR and the MSC in the TIA reference model.

- **Terminal Mobility Data Store (TMD):** The TMD maintains associated terminal data and is similar to the VLR in the TIA reference model.

- **Personal Mobility Controller (PMC):** The PMC provides control logic for user authentication, location management, alerting, and routing to users. This function is supported by the HLR and the MSC in the TIA reference model.

- **Personal Mobility Data-Store (PMD):** The PMD maintains the associated user data and is similar to the HLR in the TIA reference model.

- **OAM&P System:** The OAM&P is identical to the OS in the TIA reference model.

- **Auxiliary Services:** These represent a variety of services such as voice mail and paging, which may be provided by the PSC.

- **Interworking Function (IWF):** IWF is identical to the IWF of the TIA reference model.

- **External Networks:** These networks (i.e., wired and wireless) are not part of the described wireless network.

The following interfaces are described:

- **RP to RPT (a-interface):** This interface is identical to the U_m interface of the TIA model.

- **PSC to RPC (c-interface):** This interface is similar to the A-interface of the TIA model. If an RASC is used, the c_{rpc}, c_{rasc}, and c_r-interfaces are defined.

- **TMC to Other Elements (d-interface):** The d-interface is between the TMC and the RASC and between the PMC and the PSC. The d_d-interface is between the TMC and the TMD and between the TMC and the PMC.

- **RPI to RPC (f-interface):** The f-interface is between the Radio Port and the Radio Port Intermediary. It may or may not be internal to a radio system.

- **PSC to Auxiliary Functions (l-interface):** This is the same as the L-interface in the TIA reference model. As in the TIA model, this interface is defined by the interworking function.

- **PSC to External Networks (m-interface):** This is the same as the TIA reference model to external networks except that TIA segments this interface into type of network (A_i, D_i, M_i, and P_i).
- **PSC to other PSC (n-interface):** This interface is to other PSCs.
- **RPI to RP (p-interface):** This interface carries baseband bearer and control information, contained in the air interface, between the RPI and the RP.
- **OAM&P Systems to PSC (q-interface):** This is the same as the O-interface in the TIA reference model.
- **Terminal Adapter to Terminal Equipment (t-interface):** This interface depends on the type of equipment and is the same as the R-interface of the TIA reference model.
- **PSC to Terminal Equipment (w-interface):** This interface depends on the type of equipment and allows terminal equipment to be directly connected to the PSC. There is no equivalent in the TIA reference model.
- **OAM&P to Craft Terminal (x-interface):** This interface provides capabilities to access the operations system to display, edit, and add/delete information.
- **PSC to Auxiliary Services (y-interface):** This interface, which is the same as the X-interface of the TIA reference model, depends on the auxiliary equipment connected to the PSC.

4.3.3. Open Interfaces

Many interfaces in the reference architectures are closed and proprietary. Only the air interface (U_m- or a-interface) is fully open. However, standards groups are in the process of publishing open standards for other interfaces. Some of the open interfaces are as follows.

The **interface between the switch and base station controller** is the A-interface or c-interface. Work has progressed on several different protocols for the A-interface, based on different signaling protocols.

- **SS7-Based Signaling:** The first open interface for the switch to base station controller is based on the SS7 signaling protocol [15]. A link between the two NEs uses one 56 or 64 Kbps channel in a T1 (1.544 Mbps) connection. The underlying protocol is then a point-to-point protocol that is consistent with the SS7 protocol used for signaling between switches.
- **ISDN-Based Signaling:** A signaling protocol that enables the MSC to be either a wireless switch separate from the wireline network, or a wireline switch, has also been developed [16]. The protocol is based on ISDN BRI or PRI communications and is supported using one 56 or 64 Kbps channel in a T1 (1.544 Mbps) connection or a 16 Kbps in a 144 Kbps BRI channel.

- **Frame Relay-Based Signaling:** As an interim step toward using Asynchronous Transfer Mode (ATM) communications, and to support CDMA services, a Frame Relay Signaling System is in the process of being standardized.

The **Air Interface** (U_m interface or a-interface) is between the mobile station and the base station transceiver. This interface has traditionally been an open interface. In North America, six air interfaces have been standardized for use with cellular/PCS systems. These interfaces are:

- **Advanced Mobile Phone System (AMPS):** An analog standard [17] using Frequency Modulation for voice and Manchester Frequency Modulation (MFM) for data communications. Each AMPS channel occupies a bandwidth of 30 kHz.
- **Combined CDMA/TDMA (CCT) System:** A digital standard [18] that uses both time division multiplex and code division multiplex to assign 24 (or 32 with small cells) users to a 5.0 MHz system operating at a TDMA frame rate of 781.25 Kbps, with a 5.0 Mbps spreading code or 16 users to a 2.5 MHz system operating at a TDMA frame rate of 468.25 Kbps with a 2.5 Mbps spreading code.
- **Code Division Multiple Access (CDMA):** A digital standard [4, 19] that uses spread spectrum and code division to distinguish multiple users on the same frequency. Each CDMA channel occupies 1.25 MHz bandwidth.
- **Personal Access Communications System (PACS):** A digital standard [20] that uses time division to multiplex 7 users into a 384 Kbps channel with a bandwidth of 300 kHz.
- **Time Division Multiple Access TDMA:** A digital standard [2, 21, 22] that uses time division to multiplex three users into one 30 kHz channel. It thus replaces one AMPS channel with three TDMA users.
- **Wideband Code Division Multiple Access (W-CDMA):** A digital standard [23] similar to CDMA but using a wider bandwidth. Each W-CDMA channel occupies 5, 10, or 15 MHz bandwidth, depending on the parameters chosen for the system.

The **interface to the Public Telephone Network** is a standard and open interface. Signaling connections are based on multifrequency signaling or on SS7 signaling [24]. Voice transmission is based on μ-law Pulse Code Modulation (PCM) or standard analog voice transmission.

Inter-MSC communications: When mobile telephones move from their home system to a visited system, communication must occur between the two mobile switches. Originally, this communication was based on proprietary signaling using private lines, modem connections, or X.25. An open interface based on SS7 signaling has been developed by TIA (IS-41) [25]. The SS7 signaling can be either on a private SS7 network, linking multiple wireless service providers, or on one of the many public SS7 networks owned and operated by the local and interexchange carriers in North America.

- **Billing Interfaces** (IS-124): The standards for interfaces between switches and other NEs between carriers is based on IS-41, an application of SS7 (Signaling System no. 7) for PCS/cellular service. Major revisions were made in the existing standards, resulting in IS-41C, and a new standard IS-124 [26], referred to as Data Message Handling (DMH), was developed [16]. DMH (IS-124) moves "near real-time" messages between different places in the PCS/cellular network. The basic messages are related to call activities, when and where the call was made, how long, and so on. DMH fulfills the cellular/PCS operators' need for a way to rapidly transfer call detail records among various business entities involved in a call. DMH has critical features important to the business, such as accountability, auditability, and traceability. The uses of DMH are settlement, fraud monitoring, credit verification, credit limit monitoring, and real-time customer care in settling billing disputes. DMH provides reliable data communication protocol to support applications that should not ride on the time-critical IS-41 signaling network.

Future application of DMH could include

- Customer equipment configurations
- Short message (data component)
- E-Mail and e-mail attachments
- E911 location data

DMH has flexible message services, including:

- **Certified Delivery:** This service transfers the call detail information between NEs with a positive acknowledgment back to the sender. Thus, the sender knows when the data get through.
- **Uncertified Delivery:** This mechanism is called "send and pray," as contrasted to certified delivery. The sender does not know if the receiver got the message. This message service involves less overhead than the "certified delivery."
- **Retransmission Request:** This command is typically initiated by the receiver when it is certain some data have been lost. The receiver knows something is missing by examining the message sequence numbers. This might be used to ask for the missing information, which was originally sent using the "Uncertified Delivery."
- **Record Request:** This command is initiated by the receiving end when it needs a specific record or set of records to complete a job. For example, to complete a specific bill, an operator needs not only the network roaming bill but

also the associated long distance costs. The operator would use this message service to request that record.

- **Aggregate Delivery:** This service is similar to a batch header and trailer on an accounting tape and provides summary information on records previously sent.

- **Aggregate Request:** This is how the receiver of records asks for summary information data that were previously sent. This message service allows the receiver to verify that all data sent have been received.

- **Rate Request:** This is a request to an external service to put a dollar amount on the call, based on the call detail record.

DMH has been designed to operate under many configurations, modes, and applications, and to be scaleable, thereby enabling it to migrate throughout the network. Furthermore , it has been designed to allow operators to control the nature of the data they interexchange with other carriers and clearinghouses.

DMH is compatible with the OSI seven-layer model and is an application service in Layer 7. The OSI model allows decoupling of software between layers. Using this model allows the operator to change the network while preserving the software investment.

- **Operations interfaces:** Standard Operations interfaces are based on the Telecommunications Management Network (TMN).

4.4. REQUIREMENTS FOR PCS NETWORK MANAGEMENT

In a PCS network, users and terminals are mobile. In addition, radio resources that do not exist in a wireline network must be managed. Furthermore, the network model can be a cellular model where the radio and switching resources are owned and managed by one company, or it can be a hybrid model where a wireline company provides standard switching functions and a wireless company operates radio-specific functions. With the new hybrid mode of operation, network management requirements for PCS networks include standard wireline requirements and new requirements specific to PCS. The new PCS specific requirements are needed in the following areas [27]:

- **Management of Radio Resources:** PCS allows terminals to be connected to the network via radio links. These links must be managed independent of the ownership of the access network and switching network. The two networks may consist of multiple network service providers or one common service provider. The multiple operator environment will require interoperable management interfaces.

- **Personal Mobility Management:** Personal mobility is the association of service with an individual as opposed to a terminal. Users are no longer in a fixed location or carry a unique terminal but may be anywhere in the world using any terminal. Users can be addressed by their identifications without taking into account their current location or status. The network should recognize the originator's ID and deliver stored information that augments the user's identity. This will increase the load on management systems as users manage various decision parameters about their mobility.

- **Terminal Mobility Management:** A primary focus of PCS network is to deliver service via wireless terminals. Ultimately, these terminals may appear anywhere in the worldwide wireless network. Single or multiple users may register on a wireless terminal. The terminal associated with users is tracked, perhaps across multiple networks, in order to deliver messages or calls to the personal ID. Terminal mobility deals with the ability of a terminal to access telecommunications services from different locations while in motion. Terminal mobility management function may be integrated with the personal mobility management function or may be separate from it. Different service providers may operate their systems in a variety of ways. The PCS personal and terminal mobility function should support all modes of operation.
- **Service Mobility Management:** Service mobility is the ability to use vertical features (such as call waiting and call hold) transparent from remote locations or while in motion. As an example, the user has access to messaging service anywhere, anytime. The ability to specify a feature is provided via a user interface that is flexible enough to support a number of input formats and media. A user can specify addressing in a simple, consistent way, no matter where he or she is.

4.5. MANAGEMENT GOALS FOR PCS NETWORKS

Early cellular systems were purchased from equipment vendors as complete and proprietary systems. Interfaces between NEs were proprietary to each vendor, and NEs from one vendor did not work with NEs from another. Included were management systems tailored to vendor's hardware and software. Now, both cellular and PCS networks are migrating to a mixed-vendor environment with open interfaces. Thus, a service provider can buy switching equipment from one vendor, radio equipment from another, and network management systems from a third. In this new environment, specific goals for network management of hardware and software are necessary. Network management goals for PCS are:

- **Operation in a mixed-vendor environment:** Network management hardware, software, and data communications must support NEs from different vendors. It should support integration of equipment from different manufacturers by using clearly defined interworking protocols.
- **Use of existing resources:** Existing wireless companies should be able to manage their PCS network with minimum additions of new data communications, computing platforms, and so on. If possible, minimum additions

should be made to existing distribution networks to implement the required control or reporting system.

- **Support for multiple and interconnected systems:** When a service provider has multiple systems deployed, the different management systems should interact effectively to deliver service by sharing a common view of the network.

- **Support to share information across multiple-service providers:** Multiple-service providers must interact to share information on billing records, security data, subscriber profiles, and so forth, and to share call processing, that is, intersystem handoffs. Network management systems should therefore, allow flexible telecommunications management relationships among multiple-service and network providers of PCS.

- **Support for common solutions between end-users and service providers:** Many services require a joint relationship between the service provider and end-user. End-to-end data communication is the most common example of the need for a joint relationship. When service providers and end-users are operating in this mode, their management functions should be interconnected. Solutions should allow flexible telecommunications management relationships between service providers and end-users. For example, the end-users should be able to have bandwidth on demand.

- **Transparency:** A PCS network management system must be transparent to the technology used in network implementation.

- **Flexibility:** The network management system must be flexible to allow for evolution in PCS network functions and services.

- **Modularity:** The PCS network management system must be modular, so that regardless of what network size will be reached or where the control/knowledge will reside, the management functionality can support all management aspects.

- **Fail-safe:** Neither equipment failure nor operator error should render the management system or the PCS network inoperative.

4.6. MANAGEMENT FUNCTIONS OF PCS NETWORKS

4.6.1. Accounting Management

Accounting management [28] deals with the generation and processing functions of "usage information." It includes the distributed function that measures usage of a PCS network resources by subscribers or by the network itself (e.g., for audit purpose) and manages call detail information generated during the associated call processing to produce formatted records containing the usage data. Accounting management includes billing data, information for fraud detection, and subscribers' profiles (i.e., authorization to charge, etc.). The primary aim of accounting management is to render an invoice to the subscriber that used a service. The user may be an end-user or another network. IS-124, referred to as Data Message Handling (DMH), moves "near real-time" messages between different places in the PCS/cellular network.

Call and event data are required for network management activities, including, but not limited to, the following:

1. Billing of home subscribers, either directly or via service providers for network utilization charges.
2. Settlement of accounts for traffic carried or services performed by fixed network operators.
3. Settlement of accounts with other PCS network providers for roaming traffic.
4. Statistical analysis of service usage.
5. As historical evidence in dealing with customer service and billing complaints.

To support these activities the accounting management system for PCS should support the functions listed in Table 4.2.

TABLE 4.2   ACCOUNTING MANAGEMENT FUNCTION FOR PCS

Management function	Tasks
Usage Management Functions	Usage generation; usage edits and validation of call events; usage error correction for call events; usage accumulation; in-call service request; service usage correlation; usage aggregation; usage deletion; usage distribution.
Accounting Process Functions	Usage testing; usage surveillance; management of usage stream; administration of usage data collection.
Control Functions	Tariff administration; tariff system change control; tariff class management; data generation control; partial record generation control; data transfer control; data storage control; emergency call reporting.

PCS accounting management must address the following items:

- **Distributed Collection of Usage Data:** Since many PCS services are "multi-network services" (e.g., roaming of mobile subscribers), it is difficult for a single node (such as a switch or base station controller) to generate a complete record of a call, as is done in wireline network. Usage data collection may involve multiple network nodes, possibly belonging to different service providers.
- **Improved Performance for Billing Collection and Report Generation:** Since the usage data are collected by multiple nodes in the network, the data should be transmitted to DMH in near-real time. There will be concern about data concurrency and latency restrictions as well. These issues may pose significant performance requirements.

- **Multiple Charging Strategies:** Usage data collected for billing should be flexible enough to support a variety of charging strategies. In some cases, customers may pay directly. In other cases, the provider may pay. Charges for a call may be split between calling and called parties. In the case of multiple service providers, the customer may deal with only one service provider (such as 900 service). Different legs of a PCS call may receive different charging treatment.

4.6.2. Performance Management

Performance management functions [29] deal with a continuous process of data collection about the grade of service, traffic flow and utilization of network resources, traffic engineering, and congestion control in a PCS network. Performance monitoring of PCS should be designed to measure overall service quality to detect service deterioration due to faults or to planning and provisioning errors. Performance monitoring of the air interface should be designed to detect the characteristic signal pattern before signal quality drops below an acceptable level. Performance monitoring is sensitive to sampling schemes of the monitoring signal or choice of signal parameters calculated from the raw signal data.

Monitoring PCS network behavior requires that performance data be collected and recorded by its network elements according to the schedule established by the operations systems. The purpose of collecting the data is to evaluate the operation of the network to verify that it is performing within the defined quality of service limits and to localize potential problems as early as possible. Network elements are required to generate data to support the following areas of performance evaluation:

- Traffic levels within the PCS network, including the level of both user traffic and signaling traffic.
- Verification of network configuration.
- Resource access.
- Quality of Service (e.g., delays during call setup, etc.).
- Resource availability.

Performance management phases include

- Management of the performance measurement collection process.
- Generation of performance measurement data.
- Local storage of measurement data in network elements.
- Transfer of measurement data from network element to an operations systems.
- Storage, preparation, and presentation of data to the craft.

Typical requirements of performance data to be generated by network elements of a PCS network are:

1. **Traffic measurement data:** Traffic measurements data provide information helpful in both planning and operation of the network. These include:
   - Signaling and user traffic load on the radio interface. (Examples of measurements include busy hour call attempts, number of handoffs per hour, and number of dropped calls.)
   - Usage of resources within the network nodes.
   - User activation and use of supplementary services (e.g., call waiting and so forth).

2. **Network Configuration Evaluation Data:** Once a network plan, or changes to a network plan, are implemented, it is important to evaluate the effectiveness of the plan or planned changes. Measurements required to support this activity should indicate traffic level, with particular relevance to the way traffic uses the PCS network.

3. **Resource Access Data:** For accurate evaluation of resource access, each count would need to be produced for regular time intervals across the PCS network or for a comparable part of the PCS network.

4. **Quality of Service Data:** The PCS network management integrates many PCS networks and equipment providers to achieve coherent and seamless information exchange to support QoS objectives and assist in achieving customer business objectives. The PCS network operator should establish QoS criteria and objectives in the context of service levels to be provided to customers. These objectives and expectations should be compared with those gained by monitoring the performance of the PCS network in order to uncover potential problems. ITU-T Recommendations I.350 and I.140 provide general aspects of QoS and network performance in digital networks, including ISDNs. The recommendations may be expanded to encompass QoS measures for users, with digital data services in the future. These recommendations apply to PCS networks. In ITU-T Recommendation I.350, QoS aspects are restricted to the parameters that can be directly observed and measured at the point at which the customer accesses the service. PCS network performance should be measured in terms of parameters that are useful to service providers. Network performance parameters can also be used for the PCS system design, configuration, operation, and maintenance.

5. **Resource Availability Data:** System availability should be measured by comparing the defined objectives (i.e., the availability performance activities carried out during different phases of the life cycle of the system), and the physical and administrative conditions.

4.6.3. Fault Management

Fault management [30, 31] consists of a set of functions that enable the detection and isolation of an abnormal operation of an NE to be reported to an OS. Fault management uses surveillance data collected at one or more NEs. Collection and reporting of surveillance data are controlled by a set of generic functions. Fault management functions for OS/NE interfaces deal with maintaining and examining error logs, reporting error conditions, and localizing and tracing faults by conducting diagnostic tests. Fault management of PCS network should support the following management service components:

- Alarm Surveillance: Alarm surveillance deals with managing information about service affecting performance degradation in a centralized environment. Alarm functions are used to monitor or interrogate NEs about events or conditions.

- Fault Localization: Fault localization or identification requires that the management system have capabilities for determining the unit at fault to allow the repair or replacement of the faulty unit so as to restore normal operation of the system. Such repair or replacement should be accomplished automatically by the OS, or manually. Alarm notification and test results should contain identification of the repairable/replaceable unit whenever possible. Where it is not possible to identify a specific repairable/replaceable unit, a list of potential faulty units should be provided so that additional diagnostic tests may be conducted to further localize the fault. Alarm notifications should provide identification of the NE reporting the alarms, probable cause, and optionally specific problem values to give the PCS network provider or manufacturer the information needed to localize the fault to a specific replaceable/repairable unit.

- Fault Restoration, Correction, or Recovery: Correction of faults should be either through repair of equipment, replacement of equipment, software recovery, or cessation of abnormal conditions. If the NE uses an automatic recovery action, the management system should support the NE's ability to notify the OS of changes that were made. Automatic recovery mechanisms should take system optimization into account. The craft should be able to shut down and lock resources before taking any fault management action. The OS should send a request to the NE indicating which NE resources are to be shut down. The NE locks the indicated resource immediately if it is not active. If the specified resource is active, then locking is delayed until all activities are ended. The fault management system of a PCS network should also support establishment and selection of previous versions of software and databases (i.e., backup, fallback, etc.) in order to support recovery actions.

- Testing: Periodic, scheduled, and on-demand tests should be used to detect faults in the system. On-demand tests are used to assist in localizing a fault to one or more repairable/replaceable units. On-demand tests are conducted to verify a replaceable unit before placing it into service. In-service tests should be used to gather information related to a particular PCS subscriber's activities within a network. Such data might include information related to successful and failed registrations, call attempts, handoffs, and so on. General requirements for tests and test management may be found in ITU X.737 and ITU X.745 specifications.

- System Monitoring and Fault Detection: Fault management of NEs requires the capability to support the recognition and report of abnormal conditions and events. Fault management should also address activities such as trend analysis, performance analysis, and periodic testing. Fault management requires that OSs have a consistent view of the current state and configuration of the system that is being managed. The management system should be able to request current information about the state and configuration of the system in the form of solicited reports. Local OAM&P activities should be reported to the OS. Whenever possible, an NE should generate a single notification for a single fault. An NE should support the local storage of fault information. This locally stored information should be accessible from the OS. Alarm information should be able to be turned off. NEs should support capabilities to allow the requesting and reporting of current alarm status information. This should indicate units with alarms outstanding and their severity. Fault management should have the capability to report alarms that identify equipment failures (hardware and/or software), database faults, or undesirable environmental conditions. Information about faults should be stored in the OS for statistical verification/estimation of equipment reliability. Alarm reporting functions are described in ITU X.733 specifications. They fall into four general areas: (1) generation of an event notification by a managed resource, (2) forwarding of that event notification to a management application, (3) storage (temporary) of the event record, and (4) alarm status monitoring capability.

4.6.4. Configuration Management

Configuration management [32] deals with a set of functions that are used to control, identify, and collect data to and from the NEs. It is also concerned with service provisioning and user profile management. PCS network operators need to have an overview of the entire PCS network, including, but not limited, to the following:

1. An overview of hardware, firmware, software, and combinations that are compatible with each other.
2. An overview of hardware, firmware, and software that are currently used in each network element.
3. An overview of the frequency plan for PCS networks and the frequency allocation for each cell.
4. An overview of the coverage plan and the coverage area of each cell.
5. An overview of the future panning and engineering of network resources.

Configuration management of a PCS network is concerned with monitoring and controls that relate to the operation of NEs in the system. This deals with initial installation, growth, or removal of system components.

The system modification functions are as follows:

1. **Creation of Network Elements and Resources:** Creation of a network element is used to initially set up a PCS network or to extend an already existing network. The action of creation includes a combination of installation, initialization, and introduction of newly installed equipment to the network. It also involves updates to the operations system that will control it. The creation may affect equipment, software, and data. Whenever a PCS network or parts of it are installed, the created network elements are required to be

   - Physically installed and tested and initialized (using a default configuration).
   - Logically installed by means of introduction to the network (possibly involving changes to existing network element configuration).
   - Allowed to be put into service.

   The management system should support mechanisms for user friendly identification of these elements and resources. It should be possible to associate information such as resource name, location, description, and version, with logical or physical elements.

   The sequence of physical and logical installation may vary depending on the specific PCS network operator strategy. In case the logical creation occurs before the physical creation, no related alarms should be reported to the operator.

2. **Deletion of Network Elements and Resources:** If a PCS network is found to be overequipped in a certain area, the operator may reduce equipment in order to reuse elsewhere. This situation may occur when an operator overestimates traffic in one area and underestimates in another. Deletion of a network element requires

   - Taking the affected network elements out of service.
   - Logically removing the network elements from the network (possibly involving changes to other network element configurations, for example, neighbor cell description).
   - If necessary, physical dismantling the equipment.
   - Returning other affected network elements to service.

   The sequence of logical and physical removal will not matter if the affected network elements are taken out of service prior to their removal. This will help to protect the network from error situations.

3. **Conditioning of Network Elements and Network Resources:** When a network element is to be modified, the following actions should be performed:

   - Logical removal from service.
   - Required modifications.
   - Logical returning to service.

This sequence is recommended to protect the PCS network against fault situations that may occur during the modification process. The result of modifying the network elements should be observable by the operator.

Modification of data that have a controlling effect on some of the resources could affect the resource throughput or its capacity to originate new traffic during the modification period. This should be evaluated for particular modifications, because the capacity of the network element can be reduced and thus affect the traffic. The forecast of a modification on capacity, throughput, and current activity of a resource will help the network operator to decide when a modification should be carried out.

Not all the data characterizing the PCS network will be subject to the same rate of change. Changes to the logical configuration may also need to be applied across multiple network elements.

A major aspect of configuration management is the operator's ability to monitor the current configuration of the PCS network. This is necessary to determine the operational state of the PCS network and to determine the consistency of the information request function, the information report function, and the response/report control function.

- **Information Request Function:** An operator should monitor the PCS network in order to ensure proper operation. The OS should be able to collect information for any single attribute defined in a management information base. Also, the OS should be able to collect a large amount of data from a single request by providing appropriate scope and filtering constructs in the request. On receipt of a valid request, the addressed NE in the PCS network must respond with current values of the specified data elements. This response should be immediate if the OS requests it. However, in cases where a large amount of data is involved and where the OS and NE support the capabilities, the OS may request the NE to store the data in a file and transfer it using a file transfer mechanism.

- **Information Report Function:** The NE should have the capability to report information autonomously. This will be performed when some information on the state or operation of the system has changed. For appropriate events, an NE should be able to identify the notification of an alarm and be able to indicate the severity and cause of the condition in the report. Notifications should be logged locally. Logged notifications may be requested to be transferred using the defined file transfer mechanism.

- **Response/Report Control Function:** For responses to information requests and information reports, the operator should be able to specify where and when the information should be sent. The OS and NE of the PCS network should have the capability to configure the response/reporting to meet the following requirements:
  - Information forwarding should be able to be enabled and disabled.
  - Information should be able to be forwarded to the OS as soon as it is available (e.g., alarms).

- Information should be able to be directed to any of various OSs.
- Information should be able to be logged locally by the NE and, optionally by the OS.
- Information should be retrievable from logs using appropriate filtering specifications.

4.6.5. Security Management

Security management deals with a set of functions that control and administer the integrity, confidentiality, and continuity of PCS services against security threats. These sets of functions support the application of security policies and audit trails by controlling security services and mechanisms, distributing security-related information, and reporting security-related events.

The occurrence of various security events should be recorded. Depending on the type of information, frequency of occurrence, and importance of the event, one of several mechanisms should be used to record an occurrence:

- Scanners to collect and periodically report measurement information on high-frequency or low-importance events.
- Counters, permitting the definition of threshold crossing and notification (for example, signal strength of a MS).
- Security alarms for high-importance or infrequent events.

Some security events that occur during the life of a system should be reviewed immediately and actions should be taken at once. For other events, it may be useful to review the history in order to identify patterns of failure. These data should be maintained in a log that holds security audit records. This log should be kept by either the agent or the manager.

The craft should be alerted whenever an event indicating a potential breach in security of the PCS network is detected. (For example, this may include the databases in HLR, VLR, AuC, and EIR.) The detection should be reported by an alarm notification. The security alarm report should identify the cause of the security alarm, its perceived severity, and the event that caused it. All requirements relative to store and forward of alarm notifications identified under fault management should apply to security alarms as well.

4.7. SUMMARY

In this chapter, we presented the two reference models that have been adopted for PCS networks by two North American standard bodies, TIA and T1P1. Various interfaces between network elements (NEs) of a PCS network were discussed. The need to manage a PCS network in the mixed-vendor environment was established. Two management approaches based on the Simplified Network Management Protocol (SNMP) and the Common Management Information Protocol (CMIP) were discussed, and it was suggested that CMIP-based management schemes be used. The chapter concluded by presenting high-level requirements in the five functional areas of management for a PCS network as outlined in the TMN.

References

[1] Garg, V. K., and Wilkes, J. E., *Wireless and Personal Communications Systems*, Prentice Hall Inc., Upper Saddle River, N.J., 1996.

[2] PCS-1900 MHz, IS-136 Based, Mobile Station Minimum Performance Standards, J-STD-009, PCS-1900 MHz, IS-136 Based, Base Station Minimum Performance Standards, J-STD-010, PCS-1900 MHz, IS-136 Based, Air Interface Compatibility Standards, J-STD-011.

[3] Mouly, M., and Pautet, M.-B., "The Evolution of GSM in Mobile Communications," Advanced Systems and Components, Proc. 1994 Int. Zurich Seminar on Digital Communications, Springer-Verlag, LNCS, Vol. 783.

[4] TIA IS-95, "Mobile Station-Base Station Compatibility Standard for Dual-Mode Wideband Spread Spectrum Cellular System," July 1993.

[5] ANSI T1.210 Operations, Administration, Maintenance and Provisioning (OAM&P)—Principles of Functions, Architecture and Protocols for Telecommunication Management Network (TMN) Interfaces.

[6] ITU-T Recommendations M.3010—Principles for a Telecommunication Management Network (TMN), Draft 950630, ITU—Telecommunications Standardization Sector, 1995.

[7] CCITT Recommendations M.3400—TMN Management Functions, ITU Telecommunications Standardization Sector, 1992 (under revision, December 1996).

[8] Draft CTIA Requirements for Wireless Network OAM&P Standards—CTIA OAM&P SG/95.11.28.

[9] Hayes, S., "A Standard for OAM&P of PCS Systems," *IEEE Personnel Communications Magazine*, December, 1994, Vol. 1, No. 4.

[10] T1M1.5/94-00R2—Proposed Draft Standard—Operations, Administration, Maintenance, and Provisioning (OAM&P) Interfaces Standards for PCS.

[11] Stallings, William, *SNMP, SNMPv2, and CMIP—The Practical Guide to Network Management Standards*, Addison-Wesley Publishing Co., Reading, Mass., 1993.

[12] Klerer, M., "System Management Information Modeling," *IEEE Communications Magazine*, May 1993.

[13] TIA Project PN-3169, "Personal Communications Services Network Reference Model for 1800 MHz," Proposal, June 6, 1994.

[14] Committee T1—Telecommunications, A Technical Report on Network Capabilities, Architectures, and interfaces for Personal Communications, T1 Technical Report #34, May 1994.

[15] TIA Interim Standard, IS-634, "MSC-BS Interface for Public 800 MHz."

[16] TIA TR-46, "ISDN Based A Interface (Radio System to PCSC) for 1800 MHz Personal Communications Systems," Project PN-3344, Re-Ballot Submission, August 29, 1995.

[17] TIA Interim Standard, IS-91, "Cellular System Mobile Station—Land Station Compatibility Specification."

[18] T1P1/94-089, PCS2000, A Composite CDMA/TDMA Air Interface Compatibility Standard for Personal Communications in 1.8–2.2 GHz for Licensed and Unlicensed Applications, Committee T1 Approved Trial User Standard, T1-LB-459, November 1994.

[19] T1P1/94-088, Draft American National Standard for Telecommunications—Personal Station–Base Station Compatibility Requirements for 1.8 to 2.0 GHz Code Division Multiple Access (CDMA) Personal Communications Systems, J-STD-008, November 1994.

[20] Personal Access Communications System, Air Interface Standard, J-STD-014.

[21] TIA IS-54B, "Cellular System Dual-Mode Mobile Station Base Station Compatibility Standard," April 1992.

[22] TIA IS-136.1, "800 MHz TDMA Cellular—Radio Interface—Mobile Station—Base Station Compatibility—Digital Control Channel," December 1994 and TIA IS-136.2, "800 MHz TDMA Cellular—Radio Interface—Mobile Station—Base Station Compatibility—Traffic Channels and FSK Control Channel," December 1994.

[23] PCS-1900 Air Interface, Proposed Wideband CDMA PCS Standard, J-STD-007.

[24] "Compatibility Information for Interconnection of a Wireless Service Provider and a Local Exchange Carrier Network," Bellcore Generic Requirements, GR-145-CORE, Issue 1, March 1996.

[25] TIA Interim Standard, IS-41 C, "Cellular Radio telecommunications Intersystem Operations."

[26] EIA/TIA IS-124 Cellular Radio Telecommunications Intersystem Non-signaling Data Communications (DMH).

[27] ETSI Technical specifications GSM 12 Series.

[28] Draft ANSI T1.XXX—199x, American National Standard for Telecommunications—Operations, Administration, Maintenance and Provisioning (OAM&P)—Technical Report on PCS Accounting Management Guidelines, T1M1.5/95-011R4.

[29] Draft ANSI T1.XXX—199x, American National Standard for Telecommunications—Operations, Administration, Maintenance and Provisioning (OAM&P)—Performance Management Functional Area Services and Interfaces between Operations Systems and Network Elements.

[30] ANSI T1.215 Operations, Administration, Maintenance and Provisioning (OAM&P)—Fault Management Messages for Interfaces between Operations Systems and Network Elements.

[31] ANSI T1.227 Operations, Administration, Maintenance and Provisioning (OAM&P)—Extensions to Generic Network Model for Interfaces between Operations Systems across Jurisdictional Boundaries to Support Fault Management—Trouble Management.

[32] Aidarous, Salah, and Thomas Pleryak eds. *Telecommunications Network Management into the 21st Century*, IEEE Press, 1994.

Yechiam Yemini
Columbia University
and
Geoffrey Moss
Motorola Satcom

Chapter 5

Managing Mobile Networks: From Cellular Systems to Satellite Networks

5.1. INTRODUCTION

Emerging mobile networks encompass a broad spectrum of technologies ranging from analog to digital cellular phones, cellular data networks, and wireless LANs as well as satellite networks. This chapter describes network management challenges, intrinsic to such networks, and approaches to address them. One could argue that this chapter is premature, given the embryonic stages of mobile networks technologies and their rapid changes. Many of these networks are still in the design or early deployment stage; there is insufficient operational experience to fully identify intrinsic operations management problems and solutions. On the other hand, designers and operators of mobile networks require an immediate better understanding of management technologies, even when these are in embryonic stages. We chose this second view to guide this chapter, at the risk of trying to capture a seminal and not well-understood body of knowledge.

The operations management of a mobile network can be viewed broadly in terms of three layers of activities: (1) *managing physical elements*, ranging from components of base stations and mobile switching centers to satellites, (2) *managing network functions*, ranging from connectivity to routing, and (3) *managing service functions*, ranging from delivery of quality of service (QoS) to fraud detection. Each of these layers presents unique challenges intrinsic to mobile networks. We illustrate these challenges through several examples.

At the physical element layer, mobile network operations are sensitive to interactions between network elements and their physical environment. The operations of channels connecting a mobile station to a base station depend on the structure of the geographical environment, atmospheric conditions, electromagnetic interference, and condition of base station antennas. Similarly, satellite channels depend on numerous underlying physical components, including battery power, precision of altitude control, and antenna directions. In contrast, operations management of the physical layer of stationary wire-based networks requires only minimal or no handling of their physical environment. Transmission over cables is mostly insensitive to atmospheric or electromagnetic interference conditions, and power sources are stable and dependable.

The network layer of a mobile network, too, is substantially different from that of stationary networks. Connectivity changes in stationary networks are infrequent and are typically handled through manual configuration management procedures. They often result in operational problems and disruption of services. A mobile network needs to handle rapid loss/gain of links with minimal impact on traffic. Connectivity changes are thus handled through built-in control mechanisms, with management assuming a more limited role of just configuring these mechanisms.

Similarly, the service layer of a mobile network presents novel challenges that do not arise in stationary networks. For example, in a stationary network, physical connectivity is established statically. In contrast, mobile units must acquire a channel before they can communicate, and they must maintain it through changes. This has impact on the quality of service (QoS) delivered to users. The QoS metrics of a mobile network service include connection setup time and call losses; these metrics are insignificant in stationary networks but play a central role in a mobile network. Service management must assure that the network indeed delivers the QoS expected by users.

This chapter focuses entirely on new dimensions of operations management intrinsic to mobile networks. The management of mobile networks must address generic management issues arising in any network. Network Elements (NEs) must be instrumented by management information bases (MIBs) accessed by a management protocol, and operations centers need network resources and connectivity maps. These generic management issues are considered by other chapters and books (e.g., [1, 2, 3, 4]) and so are not considered in this chapter. Rather, the entire focus is on challenges and issues that uniquely arise in the context of managing mobile systems.

This chapter is thus organized as follow: a section providing a broad overview of mobile and satellite networks, followed by three sections devoted to intrinsic management issues arising, respectively, at the physical, network, and services layers. The last section provides conclusions.

5.2. AN OVERVIEW OF MOBILE NETWORKS

This section provides a brief background review of several types of mobile networks. Readers interested in additional information on mobile networks can consult any of the following references: [7, 8, 9, 10].

5.2.1. Cellular Systems

Consider first the organization of a typical cellular telephone network, as depicted in Figure 5.1. Base stations serve as traffic concentrators from mobile stations (MS) to a mobile switching center (MSC). The MSC relays this traffic to the public switched telecommunications network (PSTN). The base station includes antennas, transceivers, communication subsystems for signal handling, channel access, multiplexing and protocol processing, and a stationary link to the MSC. MSs interact with the base station to register with the respective MSC, acquire channel allocation from the MSC when they wish to communicate, relay their transmissions to the PSTN, and switch to a new base station (handoff) when they move. These interactions involve a range of protocols that are executed by MS, base stations, the MSC, and PSTN.

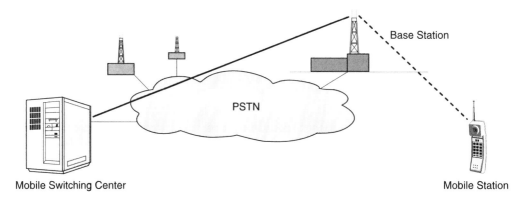

Figure 5.1 The Architecture of a Cellular System.

In general, first-generation cellular systems (e.g., the Advanced Mobile Phone System, AMPS, in the United States and its analogs in Europe, Global System for Mobile communications (GSM), and Japan) involve relatively minimal operations management complexity. These systems, built as clusters of base stations attached to an MSC, were typically small, involved homogeneous single-vendor components, and concentrated all network logic at the MSC. Vendors delivering these turnkey networks could engineer them to handle a spectrum of operational problems, leaving minimal tasks (and tools) to operations staff. Operations management focused on monitoring physical equipment or transmission problems and configuring the MSC.

Emerging cellular systems involve dramatic growth in the size and complexity of networks. They reduce cell sizes to improve bandwidth reuse efficiency, thus increasing the size of a network to provide a given coverage. They distribute a greater share of control functions to base stations, thus increasing the complexity of their operations and management. They offer multiple services, and involve greater equipment and protocol heterogeneity. These various changes result in larger and more complex cellular networks that increase the operations management complexity.

5.2.2. Satellite Networks

Traditional satellite communications systems have been based on stationary links to a geo-stationary satellite repeater. The satellite provided a "bent-pipe" relay for wide area connectivity. This section is concerned with emerging satellite networks based on Low Earth Orbit (LEO) or Medium Earth Orbit (MEO). These networks maintain efficient communications at low transmission power and small MS size by locating satellites at an altitude sufficiently close to subscribers. To accomplish such low altitudes and to support appropriate coverage, satellites must move at high speeds. Emerging LEO and MEO networks thus involve novel challenging forms of highly mobile networks.

Two distinct architectures are being used by emerging satellite networks. The first approach, used by such networks as the GlobalStar® and Odyssey® networks,[1] is based on a *"bent-pipe" communications relay* architecture. Satellites function as tall antennas, providing broad coverage of a large geographical "cell." They relay transmissions (bent-pipe) by MSs to gateways attached to the PSTN. The gateway (see Figure 5.2), forwards these transmissions to the PSTN over connections that it manages on behalf of MSs. A gateway can provide various services in addition to PSTN access; these include subscribers' registration and geo-location determination, access authorization, and call accounting. This is similar to a terrestrial cellular system. The gateways function in a role similar to that of a mobile switching center, while base-station functions are distributed between the satellites and gateways.

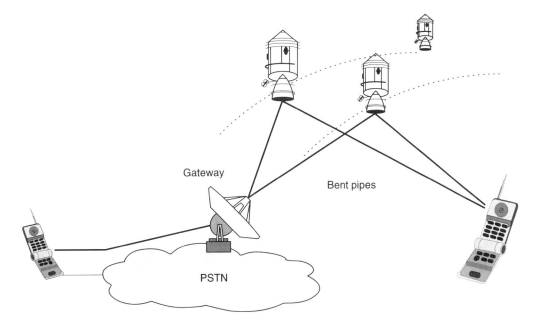

Figure 5.2 Bent-Pipe Architecture of a Satellite Network.

[1] GlobalStar® and Odyssey® are registered trademarks of the GlobalStar Corporation and TRW, respectively.

An MS acquires a channel to the gateway through satellites above it. When the subscriber–satellite link is lost, there is a need to establish a new link to support the channel; this is equivalent to the handoff problem in terrestrial systems. However, unlike terrestrial systems, an MS can access the gateway via simultaneous paths through multiple satellites. Transmission over these multiple paths may be viewed as multipath communications. This permits use of diversity techniques [9,10] to combine these multipath transmissions into a single improved reception at the ends. When a specific path is lost or a new path is gained, the MS and gateway automatically adjust to accommodate changes. Thus, handoff procedures can be entirely avoided.

A second approach to satellite networks, used by the IRIDIUM® and Teledesic® networks,[2] is based on a satellite *routing backbone* architecture. This is depicted in Figure 5.3. An MS transmits data packets over an uplink to a satellite

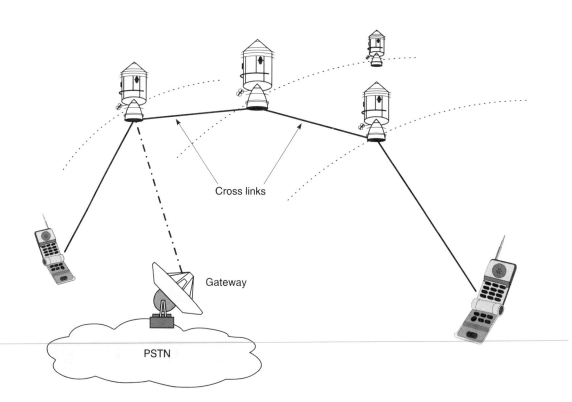

Figure 5.3 Backbone Architecture of a Satellite Network.

[2] IRIDIUM® and Teledesic® are registered trademarks of IRIDIUM LLC and Teledesic Corporation.

above it. These packets are routed by the satellites, over interconnecting cross links, to grid neighbors until they reach a destination satellite; this satellite uses a downlink to relay the packets to a terrestrial destination, whether another MS or a gateway to the PSTN. The satellite network functions in two roles. Downlink and uplink channels serve to concentrate and distribute traffic from/to subscriber units, much like the bent-pipe architecture of Figure 5.2. Cross links, however, serve as network-layer routing links and turn the satellites into a router backbone.

The architectures of Figures 5.2 and 5.3 are fundamentally different. The bent-pipe model of Figure 5.2 restricts the role of satellites to the physical transmission layer. Network layer functions such as end–end routing are supported by the gateways using the PSTN. In contrast, the routing-backbone model uses a grid of cross links to provide network layer routing functions.

Gateways are used to interconnect subscribers to the PSTN and to manage services. In principle, a single gateway can be sufficient for this. The bent-pipe architecture can be viewed as a generalization of a terrestrial cellular network where the network provides mobile links to a stationary PSTN backbone. In contrast, the satellite routing backbone model of Figure 5.2 admits the mobility not only of MS, but also of the backbone routers.

These two architectures also reflect fundamental differences in the management requirements and capabilities of the network. For example, the bent-pipe model permits management communications to a satellite only when it is located above a gateway. A network operations center (NOC) can use the PSTN to access a gateway and then use a management communication channel to poll telemetry data and events from satellites accessible to this gateway. Similarly, it can send configuration instructions to these satellites. Thus, the NOC's access to a satellite is limited to short visibility periods. This means that the NOC needs to collect, analyze, and respond to management data within very brief visibility periods. If the amount of management data that needs to be communicated during these periods is large, a significant bandwidth will be consumed by management traffic just when the satellites are to serve subscriber traffic. Furthermore, if the duration of nonvisibility by gateways is long, on-board processing resources are needed to manage problems while the satellite is not accessible. Therefore, a large number of monitoring gateways needs to be deployed to accomplish nearly continuous visibility of satellites, and on-board processing needs to be allocated to handle management and control events during nonvisibility periods.

In contrast, the router-backbone architecture maintains continuous visibility of satellites to a NOC. The NOC can use the very satellite backbone to poll telemetry data, obtain event notifications or upload configuration instructions as needed. Management communications can be scheduled to take maximum advantage of spare capacity rather than compete with subscriber units for gateway access. A router backbone architecture, however, requires management of a mobile router network. This presents technical challenges, discussed in the following sections, that do not occur in a bent-pipe architecture where network-layer functions are left to the PSTN.

Having reviewed the organization of various mobile networks, the following sections discuss the operations management challenges arising in these networks.

5.3. MANAGING THE ELEMENTS

Management is concerned with monitoring, analyzing, and controlling operations. The management of network elements is typically concerned with monitoring and configuring their operational and traffic-handling activities and with allocating their resources. Element management is typically reduced to the design of appropriate element instrumentation and its organization in a MIB. Element agents maintain an MIB with operations statistics, status, and configuration data associated with the operations of the element and report operational events. MIB objects are organized to represent resources (e.g., interfaces, protocol entities) associated with the element [2, 3, 4]. A management platform at a NOC is used to monitor these data and configure elements. The elements of a mobile network, similar to other networks, have agents that support access to MIBs and events. However, mobile networks involve new dimensions in element management that are not present in other networks. These are discussed below.

5.3.1. Managing Interactions with the Physical Environment

The operation of mobile NEs, unlike stationary networks, is significantly influenced by interactions with their physical environment. Transmissions by a cellular base station can be blocked by physical obstacles, suffer interference by electromagnetic sources, or be disrupted by a heavy rainstorm. These interactions can create a variety of operational problems that require management intervention. Management must monitor and analyze the interactions among elements and their environment. Therefore, element instrumentation must monitor not only traffic and resources but also interactions and parameters of the physical environment. For example, a cellular base station may include instrumentation to monitor electromagnetic interference or weather conditions. This instrumentation may be accessed by management software to monitor and control base station activities.

NEs in a satellite network depend on their physical environment even more critically. The physical operations of a satellite can have enormous impact on its NEs (payload). For example, the motors controlling antenna directions may be frozen, impacting the transmission layer; problems in satellite attitude control may cause complete loss of links; satellite battery power problems can cause loss of transmission or links. Moreover, these relations between network and satellite operations are reciprocal; the network can impact the satellite's operations. For example, an excessive transmission rate can deplete the satellite battery and cause its loss; damaged communications can prevent timely management access to control satellite functions. Therefore, a management system must monitor, analyze, and handle not only the NEs, but also the satellite components and their interactions with the network.

5.3.2. Managing Rapid Changes

A second distinct feature of element management in mobile networks is the handling of dynamic changes. In general, configuration changes are a primary source of operations problems and unpredictable behaviors in networks. In a stationary network, unlike a mobile network, configuration changes are infrequent, are carefully planned and executed, and are a central area addressed by operations management. Managers must reconfigure elements to handle topology changes. They must monitor the behaviors of elements to detect and handle operations problems due to these changes. In contrast, a mobile network involves topology changes that are too rapid, frequent, and random to depend on operations management intervention. Instead, elements are engineered to handle and adapt to changes automatically. Therefore, management focuses primarily on the detection, isolation, and handling of anomalies resulting from changes but not on effecting them.

We illustrate the challenges of detecting, isolating and handling problems through two examples. Consider first an element-level problem in a satellite network. Suppose one of the satellite's downlinks is experiencing a problem, leading to significant communication errors and loss. A large number of symptoms of such problems will propagate to various components of the network. For example, receivers at the other end of the link will detect physical transmission problems; link layer protocol entities will detect bit errors and loss; network or transport layer connections, layered over this link, will see loss of packets; and applications that use these connections will see performance degradation. If this downlink is used to communicate with a gateway, the gateway elements instrumentation will show these various symptom data in respective element MIBs. As the satellite moves, other gateways will show symptoms of the problem. Thus, a central challenge in detecting, isolating and handling such problems is that of correlating these diverse symptoms to diagnose their root-cause problem [5, 6].

Consider a cellular system scenario, depicted in Figure 5.4. A section of a highway is covered by two cells of base stations, attached to two separate MSCs. Cars traveling over the highway will alternate between the two cells. As they move from one cell to another, a handoff protocol will be activated to transfer their calls. The handoff is handled by protocols among the two MSCs and the respective base stations. Problems in a component along the handoff path can cause failures of the handoff and loss of calls. For example, the handoff protocol may be triggered too late by signal threshold detectors at base stations, or protocol entities may time out too quickly to abort it. These problems can cause persistent failures of handoff between the two cells and disruption of calls for customers traveling over the highway.

Detecting and isolating handoff problems is particularly challenging. A handoff failure can be best observed at the MS; however, the MSs are not directly observable by management; only base-station and MSC instrumentation can be observed. These elements may show various related events caused by the failure. For example, the base stations may have counters of lost calls, while the MSC may have counters for timeouts and aborts of handoff protocols. The instrumentation at these various

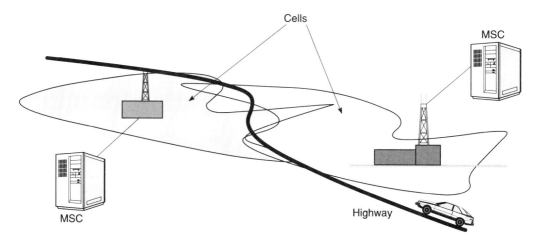

Figure 5.4 Loss of Calls Due to Handoff Problems.

elements may be extracted to detect anomalous events. A handoff failure will result in correlated events generated by various elements. None of these events may be sufficient to diagnose the problem. This is typical of end–end processes, such as handoff, that involve multiple components. Only by correlating the events of multiple elements can such a diagnosis be accomplished.

The problem can be even more difficult if its source is interactions between the mobile system and its physical environment. For example, suppose a base station is configured to trigger handoff properly during rush hour, when electromagnetic interference is high. That is, the handoff triggering thresholds at base stations anticipate interference. During times when interference is low, the handoff may be triggered too late and thus will very likely fail. A problem diagnosis system will have to monitor and correlate events at different elements with events associated with the environment, to identify the problem as a mismatched triggering configuration.

5.3.3. Managing Constrained Resources

A mobile network presents the difficult challenge of managing constrained resources under operational stress. Stationary networks are typically operated well under their capacity limits. They use flow control mechanisms to keep bandwidth resources utilized at a fraction of the capacity in order to avoid buffer overflow and loss. When a network is strained to operate near its capacity limits, it often sees thrashing performance behavior and failures. Managers eliminate this stress by allocating additional bandwidth resources and by reconfiguring flow control.

In contrast, a mobile network is characterized by limited bandwidth resources. Moreover, the revenue efficiency of a network is directly related to utilization of this limited bandwidth. Therefore, a network is typically operated near its capacity limits. In addition, satellite networks have limited battery power resources. Downlink transmissions typically present the primary demand for battery power. Again, the revenue efficiency of the network depends on the ability to allocate maximal battery power to downlink transmission and operate the network near its power capacity limits. Therefore, management of mobile networks needs to assure effective utilization of these constrained resources and to smooth operations near capacity limits.

Operations under stress have several implications for element management. First, there is a greater likelihood of failures, and thus problem management plays a more central role. Second, it is often difficult to distinguish a failure from a mere stress condition. Therefore, effective diagnosis that can correlate high-event rates to detect failures is of great importance. Third, stress at the element layer can lead to escalating problems at the network and service layers. It is necessary to configure the operations of elements to ensure that the level of stress exerted can be tolerated by these higher layers. Technologies are needed to establish and enforce operational policies to coordinate configuration of the element layer, with occurrence of problems at the network, applications, and services layers above it.

5.3.4. The Challenge of Event Correlation

The preceding discussion established the critical role of event analysis in managing mobile networks. Events signal operational anomalies that require management intervention. A fault in a hardware or software component will give rise to events indicating a failure mode. A configuration problem such as a misconfigured component will trigger a range of symptom events. A performance problem such as congestion or insufficient resources to handle loads will also result in a range of events. Thus, management of fault, configuration, and performance all depend on effective event analysis. Technologies to accomplish this are described in detail in references [5] and [6].

Mobile networks involve significantly more intensive event streams than their stationary counterparts. This is caused by several reasons. First, a mobile network involves greater randomness of operational states. As a result, built-in mechanisms can handle only a limited range of events associated with these states and leave the rest to management. Second, interactions with the physical environment can induce a spectrum of operational anomalies. Third, changes result in various operational transients that generate bursts of correlated operational events. Fourth, operations near the capacity limits of constrained resources give rise to bursts of events.

Mobile systems typically cause events to propagate more intensely among components. In general, events tend to propagate among operationally related components. Mobile systems are engineered for tight coordination among components in detecting and adapting to a broad range of operational scenarios. This creates a large range of tight operational relations leading to propagation of events among them.

The combination of rapid event generation and broad propagation presents a difficult challenge that is of central importance in managing mobile networks. How can such correlated event streams be effectively analyzed to diagnose the underlying operational problems? This problem of correlating symptom events to diagnose their root causes is known as the *event correlation*, or *alarm correlation* problem [5,6]. Event correlation in mobile networks must handle intensive event rates, random loss of events, or spurious generation of false alarms. The precision of the diagnosis can have enormous impact on network operations. In a satellite network, ineffective diagnosis can lead to potential loss of a satellite. In a cellular terrestrial network, ineffective diagnosis of a base-station problem can result in a substantial loss of calls. This can translate to a substantial business loss and frustrated customers.

A coarse computational model can illustrate the complexity of the problem. Suppose a managed component is operationally related on average to k other components. Suppose that an anomaly generates, on average, s symptom events, each propagating to all related components to cause s additional symptoms; assume that such propagation goes only m levels deep. The average number of symptoms generated by a single anomaly is thus $S = s(k^m-1)/(k-1)$. With $s = 4$, $k = 5$, $m = 3$, the average number of symptoms per anomaly can be $S = 124$. If the depth of propagation increases from $m = 3$ to $m = 4$, S increases to 624. If the rate at which an anomaly is generated in an element is given by a, and there are n elements in a system, the rate of events is given by $R = ans(k^m - 1)/(k - 1)$. With $a = 1$ anomaly/hour and $n = 250$, $s = 4$, $k = 5$, $m = 3$ as above, the rate R is approximately 10 events per second. Thus, the rates at which events are generated in a mobile system can grow rapidly with tighter relationships among components, deeper propagation of influence among multiple components, and more components and instrumentation to monitor their operations. The analysis of these events cannot be accomplished by operators and requires technologies to automate event correlation [5, 6].

5.3.5. An Architecture for Managing Mobile Networks Elements

The management problems described earlier suggest several architectural features for organizing the management of a mobile network.

5.3.5.1. How Should Mobile Elements be Instrumented? Mobile NE MIBs need to be designed to handle interactions with the physical environment, rapid changes, and operations under stress of constrained resources. Ideally, these MIBs should be standardized to permit cohesive and uniform management of similar elements by different vendors. The Cellular Digital Packet Data (CDPD) standardization provides a useful example for collaborations between cellular operators to define standards for management instrumentation.

A mobile NE must include instrumentation to reflect the potential operational impact of its physical environment. Instrumentation of the physical environment should focus on detecting critical environment events that may impact operations. For example, in instrumenting a cellular base station, one needs to monitor electromagnetic interference, multipath effects, and weather conditions. Similarly, instrumentation of a satellite needs to monitor power supply, antenna's equipment, orientation, and attitude parameters.

Management instrumentation should also monitor the statistics of critical operational changes that can trigger various protocol activities. For example, handoff protocols are triggered by threshold signal strength conditions. The statistics of such events, for example, a counter of these events, should be available to management software in order to monitor and analyze the performance of handoff protocols. Similarly, in a satellite network, the statistics of events that trigger changes in routing tables, allocation of bandwidth, or packet drop should be instrumented by MIBs.

Management instrumentation should monitor, in particular, the operations of constrained resources. It should provide a wealth of information on the behavior of these resources and detect stress conditions in their operations. Conditions such as low bandwidth availability in a mobile base station, or low battery power in a satellite, are of prime operational importance. They influence the analysis of operational events and the responses to handle them. They guide long-term planning and configuration of resources. Therefore, instrumentation of elements must provide maximally rich information on the occurrence of such resource stress conditions. In addition, instrumentation must provide configuration parameters that permit rapid response to excessive stress conditions.

5.3.5.2. How to Monitor, Analyze, and Control Mobile Elements.
Traditionally, element management has focused primarily on configuration control. Operations staff allocate resources and configure operations parameters of elements using configuration control applications. A mobile network typically automates a significant share of these configuration management functions. The focus of management is shifted to detection, isolation, and handling of operations problems, as described in previous sections. Furthermore, the complexity of problems scenario, the rate at which operations events are generated, and the growing scale of mobile networks require automation of problem management functions.

Management software must therefore monitor operations of elements, correlate events generated by their instrumentation, detect and isolate operational problems, and invoke response handlers. It must be organized to perform these functions for networks of arbitrary size and complexity, and must accommodate changes in underlying elements. Figure 5.5 below suggests a functional architecture to support such problem management. Domain managers are responsible for monitoring, detecting, isolating, and handling problems associated with a domain of network elements. A domain can consist of a single element (e.g., a base station, a satellite), a collection of related elements (e.g., base stations attached to a given MSC), or a service domain. A domain manager monitors events at elements in the domain by collecting their instrumentation. The domain manager correlates these events to detect and isolate a problem. It then invokes respective handlers to recover and correct the problem.

Domain managers can be best viewed as problem diagnosis servers. Client management applications can interact with domain managers to activate respective handlers upon notification of problems. For example, an alert client can display alerts on operator consoles, or activate pagers. A rapid-restoration client can activate a reboot process to initialize a failed element; and a problem management client can activate the trouble-ticketing system to initiate a manual handling procedure.

Domain managers, furthermore, must cooperate in accomplishing diagnosis. For example, the domain managers of two MSCs, as depicted previously in

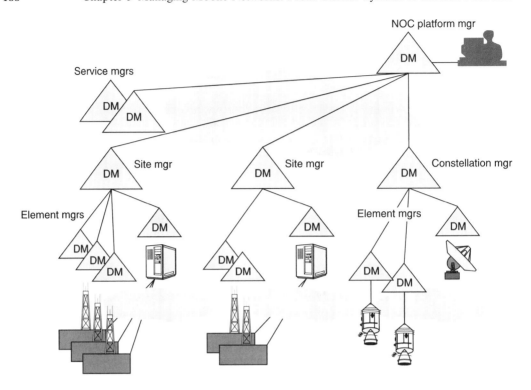

Figure 5.5 A Distributed Management Architecture for a Mobile Network.

Figure 5.4, need to correlate statistics along a handoff path to detect and isolate handoff problems to misconfigured elements. They can be organized in various hierarchical organizations to support the coordinated flow of diagnosis and handling.

The communications between a domain manager and elements in its domain are presently supported by management protocols such as SNMP or CMIP. The communications between a domain manager and a client or among domain managers require events subscription/notification protocol. This can be supported by a variety of protocols. For example, the clients can obtain event notifications using SNMP traps generated by SNMP agents embedded in the domain manager; subscriptions can be established by using SNMP SET to respective MIBs. Similarly, CMIP and CORBA can both be used to provide subscription/notifications.

5.4. MANAGING THE NETWORK LAYER

A central distinguishing feature of the network layer in a mobile network is rapid topology changes. Topology changes represent critical discontinuities in network layer operations. In a stationary network topology, changes are infrequent and involve intense manual management. In a mobile network topology, changes are frequent and are handled by built-in adaptation mechanisms. These adaptation mechanisms seek to ensure a smooth transition through changes. For example, handoff protocols seek to maintain end–end connections through physical link changes; similarly, dynamic routing mechanisms are used to reconfigure routing tables to adapt to loss of links and availability of new links. These various built-in mechanisms shift the functions of management from handling adaptation to changes to smoothing the operational discontinuities created by topology changes.

Consider, for example, an MS roaming between domains. It is necessary to adjust routing functions at the network layer to serve this MS by tracking its location. The MS must register with a visiting domain, and its services must be coordinated between the visiting domain and the home domain. These functions are provided by protocols to support roaming. These built-in protocols involve complex exchange of configuration and billing information among the visiting and home domain. A failure of these protocols translates to service or billing problems seen by the customer. A failure is often difficult to detect or handle at the protocol level; the symptoms of a failure will be observed at the visiting domain and involve both network layer and service layer events; often these symptoms require correlation with events at the home domain. Furthermore, this function requires coordination among multiple administrative domains in monitoring and analyzing these events. This function must be assigned to management.

5.4.1. Managing the Impact of Mobility on the Network Layer

In its simplest form, a mobile network involves mobile links at the physical layer with a completely stationary network layer. For example, in a cellular network, mobile links lead to a stationary backbone of MSC/base-station clusters interconnected via the PSTN. Handoff mechanisms are used to shelter network-layer end–end connection services from instantaneous link mobility at the physical layer. A similar situation exists in bent-pipe satellite networks as depicted previously in Figure 5.2. Mobile links interconnect MS to a network of stationary gateways interconnected via the PSTN. Diversity mechanisms maintain links through changes and shelter the network layer from the mobility of the physical layer. In both cases, instantaneous mobility and link changes have incidental impact on the network layer and can be managed entirely as part of handling the physical element layer.

In contrast, the satellite network architecture of Figure 5.3 involves a mobile backbone of routers. Mobility is a central aspect of the network layer operations and has enormous impact on its management. Similarly, mobile IP networks expose the network layer to mobility. In both cases the network layer must adjust its very functions—routing, flow control, security, and resource allocation—to handle mobility. Next we describe some of the implications of these adjustments to operations management.

Consider first the impact of topology changes on a mobile routing backbone operations. Topology changes require respective changes in routing tables. These changes, in turn, cause discontinuities in traffic flows. The dynamic of table changes and traffic flow changes can create forms of network turbulence, where traffic loops, gets out of sequence, or is lost.

We illustrate these possibilities through an example of routing in LEO satellite networks, as depicted in Figure 5.6. Satellites, acting as packet routers, move in four planes circling through the north and south poles as depicted in Figure 5.6*a*. Satellites are organized in belts at given latitude, consisting of four satellites each at a given plane, moving in parallel.

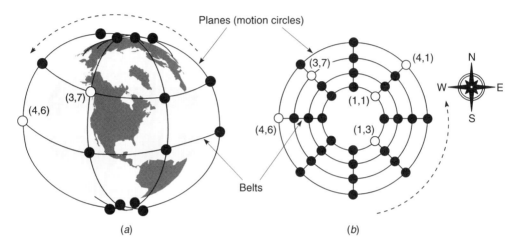

Figure 5.6 Cross Links Routing in a Satellite Network.

Figure 5.6*b* provides a simpler two-dimensional rendering of the relationships among satellites planes and belts. The planes of motion are depicted as four concentric circles. Each plane has eight satellites that are equally spaced along the circle and move counterclockwise. The eight belts are depicted as straight lines of four satellites orthogonal to the circles of motion. Neighboring satellites in a given plane have cross links along which they can exchange packets. Neighboring satellites in a given belt, except the two polar belts, also maintain cross links; these cross links are indicated by lines along belts. At the poles, the motion planes cross each other, and satellites lose their cross links along the belts.

The earth surface is divided into 26 cells: 2 polar cells are served by any of the satellites in the respective polar belt; the other 24 cells are each served by a different satellite as it sweeps above. An MS in a cell can send packets to the satellite serving the cell. The router at the satellites will forward the packets via cross links to other satellites until they reach the satellite above the destination cell. There they are transmitted over the downlinks to the destination. Routing uses cell addresses and is guided by routing tables at the satellites. A cell is addressed by a pair of numbers, indicating the respective plane (1–4) and the belt (1–8) under which it is located. The satellites move completely synchronously with each other at a very high speed. As they move, they switch between neighboring cells and must adjust their routing tables accordingly.

Suppose that a satellite above the cell (3,7) is congested with traffic destined to this cell, waiting to be transmitted over the downlinks. Suppose the satellite moves to a new position above cell (3,6) and thus switches its routing tables. The congested traffic will have to be rerouted to the new satellite above cell (3,6).

This can result, depending on traffic and routing scenarios, in several operational consequences. First, packets may be delivered out of order. Newer packets to cell (3,7) may arrive ahead of rerouted packets accumulated at the congested satellite now located at (3,6). Second, routing loops may be formed, as rerouted packets that originated with the satellite presently at (3,7) are sent back to it. Routing loops may be maintained over long terms depending on congestion levels; eventually packets are dropped when their time-to-live expires. Third, congestion levels can be maintained and even built up. The congestion at (3,6) will move to create congestion at the satellite above (3,7). In the meantime, packets originating at the cell (3,6) will be unable to proceed along the link to (3,7) and will congest the satellite at (3,6).

These various forms of traffic anomalies can be monitored by instrumentation of the packet routers in the satellites. Packet loss, high congestion levels, looping, and abrupt changes in link traffic levels will all cause threshold events to be triggered. The management of the network layer has to recognize that the cause of these events is changes in the routing tables rather than various forms of failures (e.g., loss of link, inconsistent routing tables). Furthermore, management may have to intervene and reconfigure operational parameters of the satellite routers to discard congested packets or reconfigure admission policies to reduce traffic pressures.

Similar difficult operational challenges arise in a mobile internet protocol (IP) routing backbones, presently considered by several research efforts. IP network layer functions are intimately linked with the organization of domains and connectivity. In particular, routing information protocols between domains are different from those internal to the domains. If the domain boundaries change, it is necessary to adjust the interaction of routing information protocols along the new boundaries. Similarly, firewall security requires filtering of packet flows along domain boundaries. If domain boundaries change, one must relocate dynamically firewall functions. At this time, the kind of changes required in IP network layer mechanisms to facilitate such adaptation are not known. It is unclear how management can configure and control rowing firewalls and adaptive boundary access routers.

5.5. MANAGING THE APPLICATION SERVICES LAYER

Mobile networks give rise to important and little-understood challenges in managing the application and services layer. In general, management technologies have focused on element management and only recently have begun to be concerned with management of applications and services. Management applications can access vast amounts of element data instrumented in growing element MIBs. However, it is unclear how to correlate element behaviors with their impact on the application layer.

Management of application services needs to assure high-quality end–end services. Network protocols often propagate and amplify element layer problems to cause application services problems. Even minor problems at the element layer, which are often not even observable by operations staff, can have enormous impact on the applications services layer. For example, reference [6] describes a sample scenario in which a minor interface clock problem in a physical layer element can cause the complete collapse of a data center. Monitoring and detecting service layer problems, identifying their underlying causes, and reconfiguring elements and the network layer to protect the service quality lead to substantial challenges considered by this section.

5.5.1. Monitoring the Performance of Services

Mobile networks expose the application services layer to significantly greater operational stress than stationary networks. Rapid changes in underlying layers, constrained resources, and random interference with the physical environment all contribute to this stress. These various sources of element-layer stress can significantly impact the quality of network services.

The first challenge in managing the application services layer is to monitor the quality of services delivered by the network. This leads to nontrivial challenges in instrumentation design and correlation. Where can the quality of services be observed? Consider, for example, a cellular network as depicted previously in Figure 5.1. Suppose the network takes excessive time, say over one minute, to establish a call. Customers, not used to waiting that long, will terminate and redial a call several times. Call establishment involves complex interdependent protocols between the MS, base station, MSC, and PSTN. Call setup delays can be caused due to the performance problems of any of these protocols. For example, the MS may be unable to acquire a control channel to the base station; the MSC may be congested and may be unable to complete the setup protocol with the base station; the signaling network of the PSTN may be slow to respond to the MSC in time; or these various protocols may use ill-coordinated timeouts leading to failures in synchronizing activities. Regardless of the source of the problem it can escalate rapidly. Suppose a large number of customers in a cell experience extensive call setup time. These customers will terminate and redial, leading to increasing stress on protocol processing resources and control channels, possibly exacerbating the problem and maintaining low quality of service for a long duration. Service management must, at the least, recognize the existence of a problem and, ideally, isolate the cause of the service problem, thereby activating an appropriate response. For example, thresholds for calls acceptance can be set, and a busy tone can be generated to prevent saturation of call processing resources.

Thus, the first challenge in managing services is to decide how to determine that a service problem exists in the first place. Clearly, the service qualities are best observed at end stations. In a stationary data network, the end stations can be instrumented to monitor application services properties. In a stationary telecommunications network, central offices can monitor the links to end stations. In a mobile network, management cannot depend on instrumentation at mobile stations (MS). It must thus depend on instrumentation at the base station and MSC to provide service-level operational data. For example, consider the problem of detecting excessive call setup time. Base-station instrumentation can seek to monitor congestion of control channels, or excessive delays and timeouts in completing protocol interactions with the MSC. Similarly, MSC instrumentation can seek to monitor congestion and excessive delays in completing various protocol stages in setting up calls. However, monitoring these behavior parameters of protocol processing stages leads to subtle challenges.

How can element instrumentation be used to determine excessive processing delays? How can it be used to determine congestion? Typical element MIBs collect operational data associated with protocol entities. For example, of the eleven groups of MIB II, ten are devoted to various protocol entities of a typical IP network. MIBs collect traffic processing data, associated with protocol entities, using counters of traffic events (e.g., packet arrivals, errors, discards). However, these counters of traffic events are typically insufficient to determine the average delays through a protocol processing stage. It is useful to consider the problem in terms of a queuing model of a protocol processing stage, depicted in Figure 5.7. The figure depicts a protocol processing stage and calls, or packets of data, queued waiting for this processing. Suppose a counter A collects the number of packet arrivals and a counter D collects the number of packet departures. Suppose A and D are sampled periodically. When the queue is empty, most of the time the value of A and D will be identical. If the queue is not empty, then the difference $Q = A - D$ provides the instantaneous length of the queue. Alternatively, one can instrument a gauge that measures the instantaneous length of the queue.

The instantaneous value of Q can be used to compute the average queue length Q. This value can be used to determine the expected delay through the protocol processing stage. One uses Little's formula $T = Q/\lambda$ where λ is the rate of arrivals to the queue, Q is the average queue length, and T is the average delay; the rate of arrival λ can be computed by sampling the arrival counter A at successive intervals of duration Δt and computing the average of the instantaneous rate $[A(t + \Delta t) - A(t)]/\Delta t$.

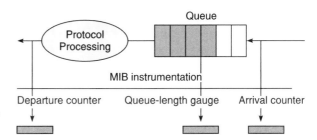

Figure 5.7 A Protocol Processing Queue and Its Instrumentation.

Unfortunately, this process of evaluating delays through protocol stages is typically difficult to implement and use. To start with, instrumentation software that updates and accesses counters typically utilizes the same processing resources as the very protocol monitored. The instrumentation software and agent software needed to access it are typically executed at lower priority as the protocol software. When a manager tries to sample the counters A and D the agent process to retrieve the data via SNMP or CMIP is typically executed after the protocol processing is completed. The requests to poll the counters will thus be served when the queue empties and then the value retrieved is $A = D$. Indeed, the interface group of MIB-II includes variables that measure the length of an interface queue. The first author conducted an experiment in which a large variety of element agents were polled for this queue length variable under various high-stress traffic levels and returned uniformly the value 0. Even if the agent processes can access the counters while the queue is not empty, typically the values of A and D will be read at different times creating significant imprecision in the estimated queue length. Even if all these problems were solved, the estimation of momentary delays requires high-intensity polling of elements, beyond what is feasible from a centralized management platform.

Current element MIBs, therefore, typically lack the instrumentation to monitor detailed performance measures as discussed earlier. A more practical approach to monitor service-level performance behavior, as in the example above, is to introduce special instrumentation at base stations and MSC to monitor application services end–end performance. For example, base stations could execute end–end dummy call setup through the MSC and measure the delays. Such loop-back tests are common at the physical layer and in IP networks are used to test reachability duration at the network layer by executing a Ping. Several recent application management instrumentation in data networks have thus pursued application-layer Ping-ing instrumentation to manage services.

In satellite networks, the problem of monitoring service performance can be substantially more difficult. Consider again the performance of call setup protocols. In a satellite network, the visiting gateway where the call is generated must coordinate the call with the home gateway. Authorization of the call and charging must be established as part of the call setup. This can involve more protocol stages to be completed across multiple network entities. The possibilities of delays and the coordination needed to analyze performance of these different stages presents a complex challenge.

Once service performance has been instrumented, it is still necessary to monitor this instrumentation and correlate it to detect the existence of a service problem. It is also necessary to trace the root cause of the problem and configure resources to correct it. This leads to event correlation problems, and distributed control problems, as discussed in the previous sections.

5.5.2. Configuration Management

Configuration management challenges arise at the element and network layer. However, at the applications service layer of a mobile network, configuration management issues present more unique and difficult challenges. Application services depend on a large number of configuration databases. To start with, end–end communications depend on mappings of names to addresses and then to routes. These mappings are maintained in various configuration databases. Access authorization, call accounting, service availability, and parameters all depend, too, on configuration databases. These heterogeneous configuration databases, spread throughout the network, must be consistent if a service is to be successfully delivered.

In a stationary network, provisioning a service involves complex transactions to modify these various databases on which the service depends. This process is relatively infrequent and occurs over periods of hours or days for a given service provisioning request. In a mobile network, automated processes configure these databases on the fly to adapt to dynamic changes. In the process of such changes, inconsistencies may be created that will disrupt services.

A central function of network management is to maximally prevent such configuration inconsistencies, and where they occur, detect and correct them rapidly. This requires technologies that can monitor, analyze and enforce consistency constraints among various configuration databases involved in service provisioning. Such technologies are pursued by a number of recent research activities.

5.6. CONCLUSIONS

This chapter illustrated a number of the management challenges arising in the context of mobile networks. While the understanding of operations management challenges and effective solutions for such networks is still in embryonic stages of development, the examples and discussions presented here suggest several conclusions.

1. Mobile networks give rise to substantially novel management problems and foci that cannot be resolved through simple extensions of management technologies and paradigms used in stationary networks.

2. In particular, the combination of rapid dynamic changes, operational impact by the physical environment, and operations near capacity limits of constrained resources lead to greater exposure to operational problems. Therefore, detection, isolation, and handling of fault, performance, and configuration problems are of central importance to management of mobile networks. Detection and isolation of problems can be accomplished by monitoring and correlating related symptom events.

3. The network layer of mobile networks can be significantly exposed to mobility. New management technologies are needed to monitor and smooth discontinuities of traffic flow caused by changes.

4. Mobile networks require a particular new focus on application services management, rather than mere element management. New technologies are required to instrument and monitor service performance. Technologies are required to assure consistency of service configuration databases throughout changes.

5. As mobile network continue to make transitions from limited, simple, and small systems to new generations of richer, more complex, and large-scale systems, operations management technologies will play an increasingly critical role in assuring effective operations of these networks.

References

[1] Aidarous, S., and Plevyak, T., eds., *Network Management into the 21st Century*, IEEE Press, 1994.

[2] Rose, M. T., and McCloghrie, K. Z., *How to Manage Your Network Using SNMP: The Network Management Practicum*, Prentice-Hall, 1995.

[3] Stallings, W., *SNMP, SNMPv2, and RMON. Practical Network Management*, 2nd ed., Addison-Wesley, 1996.

[4] Yemini, Y., "A Comparative Critical Survey of Network Management Protocol Standards," in *Network Management into the 21st Century*, S. Aidarous and T. Plevyak, eds., IEEE Press, 1994.

[5] Yemini, Y., Kliger, S., Yemini, S., Mozes, E., and Ohsie, D., "Fast and Robust Event Correlation," *IEEE Communications*, May 1996.

[6] http://www.smarts.com/products/paper_intro.html

[7] Rappaport, Theodore S., *Wireless Communications Principles and Practice*, Prentice Hall, 1996.

[8] Steele, R., ed., *Mobile Radio Communications*, IEEE Press, 1994.

[9] Boucher, N., *Cellular Radio Handbook*, Quantum Publishing, c. 1991.

[10] Yacoub, M. D., *Foundations of Mobile Radio Engineering*, CRC Press, 1993.

Chapter 6

John Brouse
Jones Intercable
and
Mohamed Beshir
Nortel

Management of CATV Networks

6.1. INTRODUCTION

This chapter presents the management functions and issues related to broadband cable television (CATV) networks. The reader is assumed to have neither prior experience with nor practical knowledge of these networks. To establish a reference point, network management is defined as the body of processes, practices, and systems employed by CATV operators to manage the cable network. This definition will be maintained throughout the chapter. We start with an overview of the subject matter to establish a general understanding of the material context, and then break down the component parts into detailed fundamental building blocks, concluding by reconstructing the overview perspective. Accordingly, we begin with a brief history of the CATV industry and then move quickly to cover current-day networks.

The majority of this chapter deals with the variety of existing CATV networks and their characteristics; a small portion of the chapter presents a forecast of future network characteristics and the demands these place on network management. To ease understanding of the topic at hand, a simple network model is used where the physical network is considered to be composed of three major components according to their basic functions (see Figure 6.1). To further enhance understanding, it is assumed that the Customer Premise Equipment (CPE) is a component part of the drop. In order to develop an appreciation for the interrelationships of the various network components, each portion of the network is covered in detail.

Figure 6.1 Physical Network Layout (Simple Model).

6.2. INDUSTRY HISTORY

During the early days of the television broadcast industry, reception of the television signal was limited primarily to the local community in which the broadcaster operated. As the industry developed, residents in rural and suburban areas desired service. Because their homes were much farther away from the transmitting antennas than those residents living in town, their television signal reception was at a much lower level. To be able to adequately receive these new television services, residents in rural and suburban areas began using more sophisticated receiving antennas, rotors, and amplifiers. As the distance from the broadcaster's antenna to the residence increased, the signals decreased to an unusable level. Also, in mountainous communities, residents living in the valleys either could not get sufficient signal levels or were unable to pick up the television signals altogether because the mountains blocked those signals (see Figure 6.2). In 1948, somewhere in western Pennsylvania (and about the same time in Oregon), a television sales and repair shop owner was having difficulty selling television sets because reception in the valley was so poor no one was buying them. In an effort to promote television sales, the store owner placed an antenna on the top of the nearby mountain and, using a cable, brought the signal down to the residences in the valley community. In essence, the entire community used that mountaintop antenna as a means for television reception, and hence the Community Antenna Television (CATV) system was born.

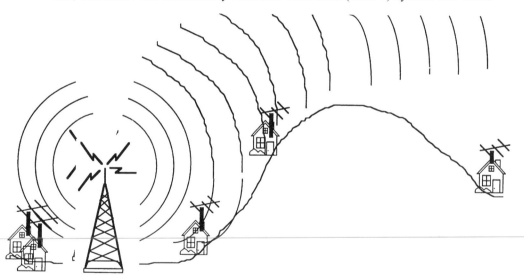

Figure 6.2 Signals Blocked by Mountains.

In addition to low-signal levels, other impairments observed in these rural communities included multipath reception and co-channel interference. Multipath results from receiving the same signal twice. The original signal is received directly, and the secondary signal is received after it has been reflected by some object (see Figure 6.3). Co-channel results when the television receives two stations operating on the same frequency (or channel). While the Federal Communications

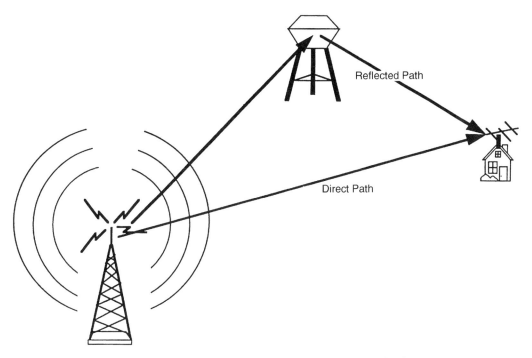

Figure 6.3 Secondary Signals Reception Through Reflection.

Commission has taken great care to ensure that television stations assigned the same channel are geographically located far apart, some meteorological conditions disturb the normal propagation of signals, which can result in a distant station being received. Early attempts to solve difficulties of low signals, multipath, and co-channel interference in these rural communities involved several methods for collecting and controlling these signals in a centralized facility, which became known as the headend. Usually, the headend included a steel tower of sufficient height with high gain and highly directive receiving antennas which were pointed at the television station's transmitting antennas. Normally, the higher the receiving antenna, the better the signal reception until the receiving antenna is at the same height as the transmitting antenna. Of course, in mountainous areas, a simple wooden pole was placed on the mountaintop, and the antenna(s) aimed appropriately. These highly directive receiving antennas usually minimized multipath and co-channel interference problems.

Early cable systems were nothing more than receiving antennas, transmission cables, broadband amplifiers, connectors and taps, and the television set (see Figure 6.4). This simple cable system connecting the received television broadcast signals, from the headend to the subscriber, was made possible by war surplus coaxial cabling and broadband amplifiers made with vacuum tubes and powered directly from the utility power company's 120VAC. The subscriber was connected directly to a single cable-amplifier system so that as the signal propagated down the cable a portion of the signal was tapped off and sent to the subscriber's television set. The

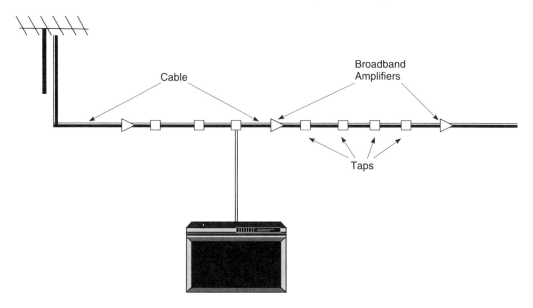

Figure 6.4 Simple Single-Cable Amplifier System.

remaining signal continued down the cable and, eventually, was attenuated enough that it needed to be amplified (see Figure 6.5). These cable-amplifier cascades were very sensitive to variations in ambient temperature and required constant adjustment by technicians. Because the attenuation characteristics of the coaxial cable rose sharply as the television channel (carrier frequency) increased, early cable systems limited carriage to channels 2 through 6. Television broadcast stations operating in the ultrahigh frequency (UHF) band (channels 7 through 13) were converted to channels in the 2–6 television station band at the headend facility. (Table 6.1 illustrates the frequency relationships between television broadcast stations and cable television systems.)

Figure 6.5 Cable Amplifier Cascades.

Through the 1950s, more cable systems were being built, surplus coaxial cable became scarce, and the need to carry more channels rose. In response, manufacturers developed improved amplifiers and cables. The cable-amplifier cascade network design was abandoned for the new trunk-feeder design. This new network design identified two distinct functions. The first function, the trunk, transported the television signals from the headend to the extremities of the system; the second function, the feeder, isolated the subscriber from the trunk and bridged the signals from the trunk to the subscriber. As the transistor developed, cable-amplifier performance improved and consumed less power; this allowed the cable plant itself to carry

TABLE 6.1   TELEVISION CHANNEL-FREQUENCY CHART

Broadcast television		Cable television	
Channel	Frequency (MHz)	Channel	Frequency (MHz)
2	55	2	55
3	61	3	61
4	67	4	67
5	77	5	77
6	83	6	83
7	175	7	175
8	181	8	181
9	187	9	187
10	193	10	193
11	199	11	199
12	205	12	205
13	211	13	211
14	471	14(A)	121
15	477	15(B)	127
16	483	16(C)	133
17	489	17(D)	139
18	495	18(E)	145
19	501	19(F)	151
20	507	20(G)	157
21	513	21(H)	163
22	519	22(I)	169
23	525	23(J)	217
24	531	24(K)	223
25	537	25(L)	229
26	543	26(M)	235
27	549	27(N)	241
28	555	28(O)	247
29	561	29(P)	253
30	567	30(Q)	259
31	573	31(R)	265
32	579	32(S)	271
33	585	33(T)	277
34	591	34(U)	283
35	597	35(V)	289
36	603	36(W)	295
37	Not Used	37(AA)	301
38	615	38(BB)	307
39	621	39(CC)	313
40	627	40(DD)	319
41	633	41(EE)	325

electrical power for powering the amplifiers. A step-down transformer fed low-voltage (30VAC) power to the cable trunk through a power inserter. The power inserter was a power block that isolated the television signals from the power supply (the step-down transformer connected to utility power) and allowed the 30VAC electrical power to be inserted on the coaxial cable. Each amplifier was powered by taking a portion of power from the coax cable. Amplifiers on the trunk portion of the network were called trunk amplifiers; amplifiers on the feeder portion of the network extended the useful reach of the feeder cable system and were, therefore, termed *line extender amplifiers*. Power could be bridged out of the trunk amplifiers to the line extenders. To limit line voltage drop problems, each power supply supplied only a segment of the network. Power blocks were used to isolate the segments.

During the 1960s, significant advances were made in the design of cable connectors. The main problem, heretofore, was keeping the connectors tight and weatherproof in order to prevent moisture from deteriorating the cable, limit reflections of signals and signal losses, and reduce power interruptions. To get the signals from the feeder cable to the subscriber's home, a pressure tap was used. A hole was cored in the outer aluminum sheath of the cable; the pressure tap center pin was centered in the hole and joined to the cable's center conductor. Self-tapping threads and a clamp provided the pressure to clamp the tap to the cable. Also during this period, improved construction practices and manufactured product improvements allowed cable operators to provide better pictures more reliably. In addition, television channels 7 through 13 in the upper portion of the very high frequency (VHF) range could be used. This resulted in a cable system capable of offering a total of 12 channels. Still more advances were made; directional couplers and signal splitters improved and led to development of the subscriber multiport tap which replaced the pressure tap.

Throughout the 1960s and 1970s, improvements in amplifier performance and cost were realized. Prior amplifiers created distortion products from channels 2 through 6, which fell in the midband frequency range; state-of-the-art amplifiers eliminated these and allowed programming to be carried in this region of the spectrum. Some cable system operators also carried local FM radio programming in the relatively clean 88MHz–108MHz portion of the midband; still from 120MHz–174MHz, an additional nine television signals could be carried. This brought the total channel-carrying capacity of cable systems to 21. However, the subscriber's television set at that time could only tune to the television broadcasters VHF channels (2–13) and their ultrahigh frequency (UHF) channels 14–83. Note that the broadcasters' VHF channels and the cable operators' channels 2–13 have identical carrier frequencies; however, the broadcasters' UHF channels (14–83) begin at 471MHz, which was beyond the capabilities of the cable system. Also during this time, the number of UHF television broadcasters increased. Since cable systems could only carry up through channel 13 (216MHz), cable operators converted UHF television signals at the headend and carried these signals in the midband (cable channels 14–21 in the 120MHz–174MHz region) to the subscriber's television set. Since the subscriber's television set could not tune to these frequencies, the cable operator placed a block-up converter before the television set and converted these midband signals to the television's UHF tuner range. For example, cable channels 14–22 were up converted to the television's UHF channels 30–38. Figure 6.6 illustrates the block up converter connection at the subscriber's television.

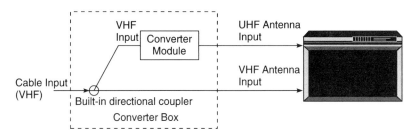

Figure 6.6 Block Up Converter Connection.

By the late 1970s and early 1980s, made-for-cable programming such as Home Box Office (HBO) and Cable News Network (CNN) became available through satellite delivery systems. Since the majority of satellite-delivered programs were premium or pay channels, some method for securing them from unauthorized viewing needed to be used. A sharp-notch filter made up of resistors, capacitors, and inductors was made and housed in a metal weather-tight enclosure. This device was connected in line with the subscriber's drop wire (usually at the tap) and effectively removed or trapped out the unpaid for channel from the subscriber's home. This was known as a *negative trap*. Defeating this mechanical security device only required removal of the trap from the line. Additional satellite programmers entered the scene, and soon more than 24 channel satellite systems were operational. This caused many cable system operators to make plans to expand their network's channel-carrying capacity either through a system upgrade or rebuild. Of course, improvements in coaxial cable and amplifier performance continued, and new cable television networks were designed and built to carry up to 30 channels (270MHz), 35 channels (300MHz), 40 channels (330MHz), 52 channels (400MHz), 60 channels (450MHz), or even 80 channels (550MHz). While this led to a dramatic increase in the number of television channels carried on the cable system, most television sets still only tuned to channels 2–13 and the VHF stations. Certainly, a number of new television sets were able to tune to some of the newer cable channels and were called cable-ready; however, the vast majority of televisions could not. So the cable system operator deployed the converter to solve this problem. The converter was designed to receive the cable signal and convert it to a fixed channel (usually channel 3 or 4) and feed this to the television receiver. Figure 6.7 illustrates several methods for connecting the converter.

As the 1980s continued, trapping of the premium and pay services was becoming unwieldy because traps had to be stacked (placed end-to-end) to trap out the many premium services from an unauthorized subscriber. To solve this problem, methods of signal scrambling were deployed. One approach was to place an interfering signal very close to the video carrier of the pay-service signal. This renders the channel unwatchable. To authorize service to the subscriber, a sharp-notch filter (or

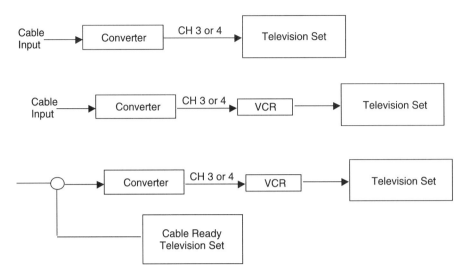

Figure 6.7 Various Methods for Converter Connections.

trap) placed on the subscriber's tap port removes the interfering signal and the channel is viewable. This trapping method is known as a positive trap. It should be noted that negative traps remove service from an unauthorized subscriber and that a positive trap provides the premium or pay service to an authorized subscriber. Other methods of scrambling used video processing equipment at the headend, which removed or suppressed the television signals' synchronizing pulses. No television receiver could view a service that had these pulses removed. The signal-restoring circuits were built into the subscriber's converter and, when service was authorized, these circuits would restore the suppressed synchronizing signals. In the latter 1980s, as the personal computer developed, the tunable subscriber converter was made addressable and included a built-in descrambler. With this type of service-security device, a subscriber could call the cable office and request a change of service; the customer service representative at the cable office could simply enter the order in the computer and, through a computer-to-cable modem, instruct the converter to authorize the desired service. No visit to the subscriber's home would be needed, and the change could be effected in minutes rather than days.

One final form of signal-security that cable system operators have deployed is addressable interdiction. The interdiction system uses military communications disruption (jamming) technology and places interfering signals over the premium or pay services. Unlike the positive trapping or converter scrambler systems, the interdiction system originates the interfering carrier at the subscriber's tap rather than at the headend. In other respects, interdiction is similar to the addressable converter system except that it removes the requirement for a converter in the home for those television sets that are cable compatible. That is, they are able to tune to all the cable channels offered by the local cable operator. This allows the subscriber full access to all the features built into the various consumer electronics equipment such as picture-in-picture and recording one program while watching a second, or third, program.

6.3. THE MODERN CATV NETWORK

Much like earlier cable television networks, modern CATV systems can be broken down into three primary subsystems according to their specific functions (see Figure 6.8); these subsystems are the headend, outside plant, and drop. The headend functions primarily as a centralized control point where all signals are collected, processed, and combined. The function of the outside plant (also called the *distribution network*) is to transport the combined signals from the headend to the customer and, in cases where the cable operator maintains a two-way plant, to transport signals from the customer or the distribution network's status monitoring and control devices back to the headend. The drop subsystem, which for purposes of this discussion includes the customer premise wiring and equipment, provides customer-authorized access to the cable operator's services.

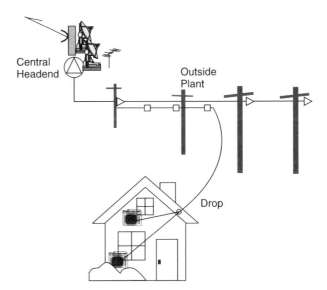

Figure 6.8 CATV Subsystems.

6.3.1. The Headend

The heart and soul of modern cable television system operations is the cable headend. It is here that essentially all the signals are collected, processed, and readied for distribution. The headend also serves as the central control point for automated network monitoring and control functions as well as the collection point for automated customer-requested service authorizations. The headend typically consists of a climate-controlled building with adjoining real estate housing an antenna farm (see Figure 6.9). The antenna farm is designed for the reception of over-the-air television broadcast and satellite delivered signals. A steel tower is normally constructed for the placement of antennas designed to the specific transmitting frequencies or group

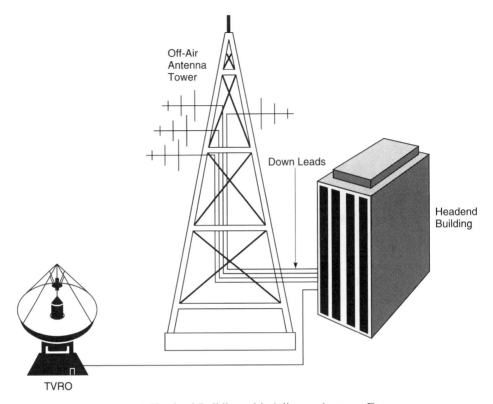

Figure 6.9 Headend Building with Adjacent Antenna Farm.

of frequencies (channels or group of channels) and aimed toward the local broadcast television stations' transmitters. In addition to these antennas, usually a broadband (VHF/UHF) antenna fitted with a rotor is placed on the tower as an emergency or part-time use antenna. Occasionally, some limited locally originated programming may be transported to the headend from a remote location via microwave; this requires the addition of a microwave receive antenna to the antenna tower. Placement of the antennas is critical; proper spacing must be maintained so that the tuned elements of any one antenna do not interfere with the reception pattern of any other antenna. The received signals from these antennas are transported into the headend building by coaxial cables called *down leads*. Over-the-air signal reception levels are a function of transmitted power, distance between the transmit and receive antennas, gain of the receiving antenna, and antenna alignment.

Assuming the receiving antennas were selected and positioned properly, maximum received power should be realized. This does not, however, assure an adequate received signal level for proper signal processing. The transmitted power level could be low, the distance between the transmitting and receiving antennas could be significant, or both transmitted power level and physical distance factors could be contributors of low received power. To make these low signal levels useful, a preamplifier is required and is placed either on the tower very near the antenna or inside the headend building. Signal performance calculations and ease of maintenance will determine where the preamp is positioned. The preamp is an active device and is powered from the headend via the coaxial cable. Once over-the-air television broadcast signals are inside the headend building, they are either demodulated and remodulated or heterodyne processed. The preferred method is to de-mod and re-mod the signals since this method provides the best technical performance and enhances operational flexibility. The de-modulator changes the received television signal into its baseband video and audio components and enhances control over these signals. These signals require eventual re-modulation to put them into the format of a television signal on a specified cable channel. Unlike the de-mod/re-mod method, the heterodyne signal processor takes the existing over-the-air signal and converts the signal in its entirety, changing its frequency to that of the desired cable channel. For example, a UHF television signal broadcast on channel 14 (transmitted over-the-air on 471.25MHz) is converted to cable channel 14 (transmitted over cable on 121.25MHz) by the signal processor. Both methods are used. However, since the heterodyne signal processing method is less expensive, cable operators usually selected it over the de-mod/re-mod method.

Made-for-cable programming such as CNN or HBO, as well as television broadcasting superstations like WWOR from Chicago, are delivered to the cable headends via satellites and require placement of television receive-only earth station antennas (TVROs) on the antenna farm property. Like over-the-air signals, TVRO signals are transported into the headend building over coaxial cables. However, TVRO signals are in the 4-GHz range, which is significantly above the UHF portion of the spectrum and exhibits significantly greater attenuation over the coax cable. To move the TVRO signals to a less lossy frequency range of the coaxial cables, low-noise block converters (LNBCs) are used and placed at the TVRO. The LNBs convert the 3720MHz–4180MHz TVRO signals to 950MHz–1450MHz signals and put them onto the coax. The LNBCs are active devices and, like over-the-air antenna preamps, are powered via the coaxial cable from the headend. Once inside the headend building, the block-down converted satellite signals are routed to either a satellite receiver, then a decoder (if the satellite service is scrambled by the service originator), or an integrated satellite receiver (IRD), which combines a decoder with the satellite receiver in one package. The outputs of the satellite receiver and decoder are baseband video and audio signals that must be modulated; that is, they must be put into the proper television signal format on the desired cable channel.

In addition to over-the-air television broadcast signals and satellite video services, many cable operators carry frequency-modulated (FM) stereo radio services in their cable networks. Some operators may simply put an FM antenna on their tower, send the signals into the headend via a coaxial down lead, and do rudimentary processing of the entire FM band. Most operators, however, individually process each FM channel with a tuned FM signal processor. The FM signals are then combined and filtered by a bandpass filter covering the 88MHz–108MHz spectrum.

Two other sources of video-related signals must be covered. These are commercial advertisement insertion and locally generated programming (also known as local origination, or LO). Many satellite program suppliers make vacant advertising slots available to cable operators to locally insert commercial ads. Each ad spot designated as available (avails) is marked at the beginning and end by start and stop signals. These start and stop signals are usually dual-tone multifrequency (DTMF) tones (called cue tones). Historically, these cue tones have been audible and were annoying to the cable subscriber; they are now being replaced with inaudible tones. Older commercial insertion equipment required that commercials be placed on a magnetic tape in time sequence. Newer systems allow the ad to be placed anywhere on the tape. Of course, each ad must be identified in order that the system can call it up during its assigned spot. This method is called random access or spot random access. The most current ad insertion systems use some type of controlling computer. With this system, the schedule is entered, the tape machines are controlled, and logs of the ads are recorded under the computer's control. The controlling computer gets program signaling from a network interface unit (NIU), selects the next available VCR, and switches the input to the appropriate television modulator from the satellite feed to the VCR output. Accordingly, the ad insertion equipment is placed after the satellite receiving equipment and before the cable modulator (see Figure 6.10).

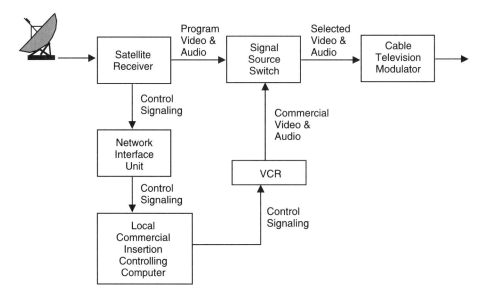

Figure 6.10 Locally Generated Commercial Insertion.

Like the cable operators' inserted commercials, LO channels are available only to cable subscribers and cannot be acquired by others either through an antenna or a satellite system. The character of an LO channel varies from the simplest character generator-based community bulletin boards to sophisticated live news or special interest programs. In the context of this chapter, considerations within the headend are similar to over-the-air television broadcast and direct feeds; that is, the signals are brought into the headend in baseband video and audio form and sent to a television's modulator which puts the signals into the proper format for distribution through the cable network.

Many cable programs are being formatted for stereo audio or a second language. In addition, many television stations broadcast the audio portion as multi-channel television sound (MTS), which provides broadcast stereo television (BTSC) as well as the second audio program (SAP). To provide this stereo and second-language audio service, the cable operator uses a BTSC encoder between the signal source and the television modulator (see Figure 6.11). One last piece of processing equipment that may be used is the addressable scrambler and is inserted either before the modulator or between sections of the modulator. The position of the scrambler is a function of the scrambling scheme. For example, signal flow for baseband scrambled signals goes from the satellite receiver/decoder to the addressable scrambler and then to the modulator. In the case of IF (intermediate frequency) scrambled signals, the signal flow is from the satellite receiver/decoder to the initial sections of the modulator, to the addressable scrambler, and then back to the final stages of the modulator. Of course, these scramblers cannot stand alone and must also be connected to a headend controller, which in turn interfaces with the customer billing system.

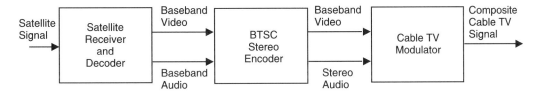

Figure 6.11 BTSC Encoder for Stereo and Second-Language Services.

Once all the acquired and locally generated signals are processed, formatted onto their respective cable channels, and scrambled if need be, they are then combined (through a combining network) into one frequency division multiplexed (FDM) signal. Other signals may also be inserted at the combining network. These include control signals for outside plant devices as well as customer premise equipment. In the case of older cable networks, the multiplexed signals are then put onto a coaxial cable and transition to the distribution network (the outside plant subsystem). The newer cable networks divide (not to be mistaken for demultiplexing) this FDM signal into multiple feeds. Each feed is then combined with unique signals and sent to fiber optic transmitters. The output of each optical transmitter may either be split multiple ways to feed multiple optical node in the outside plant or connected directly to an optical fiber that connects to a single optical node in the outside plant.

Before leaving the headend, some mention must be made of the reverse or return portion of the headend. Cable television return systems within the headend facility are used to acquire LO programming from beyond the headend facility, to provide network status monitoring and control signaling, to acquire information and data from customer premise equipment, and to provide the second channel for duplex systems. In the context of two-way cable networks, it's only today's state-of-the-art networks that deploy a significant amount of fiber optic technology and offer insight into the return portion of the headend. Suffice it to say that the return systems are similar to the forward system's objectives; that is, to acquire the signals, distribute them appropriately, and process them. Thus, the return feeds enter the headend via an optical fiber and are fed into an optical receiver. The RF output of the optical receiver may be combined with several other reverse optical receiver outputs before the signal is split multiple times. Or each reverse optical receiver's RF output may be immediately split multiple times according to the number of signals being returned and their uses. The outputs from the signal splitters are grouped by their functions and distributed to the associated processing equipment. An example of this multiple splitting and combining according to the signals functions is illustrated in Figure 6.12.

By now it should be evident that a thorough understanding of the headend and its complexity is required so that it can be properly managed, not only from the perspective of its day-to-day operations but also for ease of evolution planning. Management systems associated with the headend will be presented in detail in a later section of this chapter.

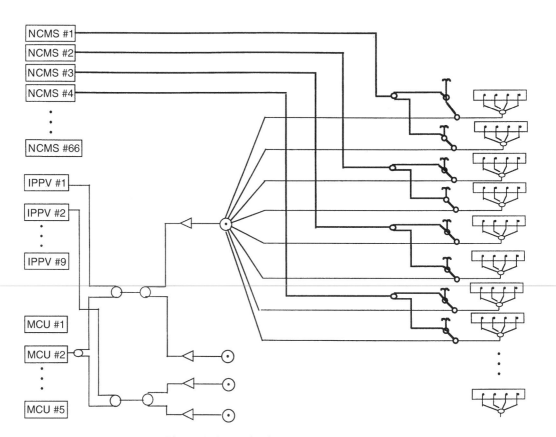

Figure 6.12 Headend Reverse RF Network.

6.3.2. Outside Plant

If the headend is the heart of a cable television system, then the outside plant represents the arteries and veins of the system. That is, the outside plant connects the headend to the community being served. Feeding directly out of the headend can be either coaxial cables, fiber optic cables, microwave transmission, or a combination of the three. Regardless of the initial medium out of the headend, eventually, all signals are placed on a coax-based network prior to arriving at the customer's premise. Therefore, the coaxial portion of the outside plant is of primary importance. In designing the coax plant, the unity gain concept is used. Consider first that the coaxial cable is a passive device and attenuates cable television's RF signals as they travel along the cable. The rate of attenuation is a function of cable size and signal frequency (see Table 6.2). Other passive devices (which we will discuss later) also attenuate the RF signals. When the cable television signals are first placed on the coax cable, they are at some power level. In the cable television industry, power levels are normally stated in dBmV referenced at 75ohm impedance. As the cable television signals travel the length of the coaxial cable and through the various passive devices, they become too weak to be usable and require amplification. In designing the coaxial network, engineers choose a fixed-gain amplifier and place the amplifier in the network at the point were the preceding cable and passive losses equal the amplifier gain.

TABLE 6.2   ATTENUATION OF COAXIAL CABLES (AIR DIELECTRIC)

Frequency (MHz)	.500″ dB/100 Ft	.850″ dB/100 Ft	.750″ dB/100 Ft	.1000″ dB/100 Ft
5	0.14	0.11	0.1	0.07
30	0.35	0.28	0.25	0.18
55	0.48	0.38	0.34	0.24
10	0.65	0.51	0.46	0.33
175	0.85	0.69	0.66	0.45
250	1.03	0.83	0.72	0.54
300	1.14	0.91	0.79	0.6
325	1.19	0.95	0.83	0.62
350	1.23	0.99	0.86	0.65
375	1.28	1.02	0.88	0.67
400	1.32	1.06	0.91	0.7
450	1.4	1.13	0.97	0.74
500	1.48	1.19	1.03	0.78
550	1.55	1.25	1.08	0.82
600	1.63	1.34	1.11	0.87
700	1.77	1.45	1.21	0.94
800	1.91	1.58	1.3	1.02
900	2.03	1.67	1.39	1.09
1000	2.15	1.77	1.47	1.16

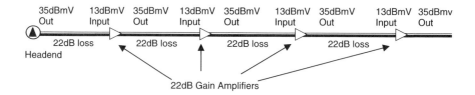

Figure 6.13 Unit Gain Concept.

Figure 6.13 illustrates this concept. Notice that the amplifiers have the same input levels and the same output levels. The unity gain concept then considers the gain of the amplifier equal to the losses preceding the amplifier. The coaxial network usually begins with one coaxial cable and, using a combination of directional couplers and amplifiers with multiple RF output ports, quickly becomes a branching network of cables (see Figure 6.14).

The classic cable television network contains two levels of design. The first level is the trunk whose primary function is to transport the cable signals from the headend to a community for distribution to the subscribers. The second level, normally called the feeder network, is connected to the trunk through a bridging amplifier and

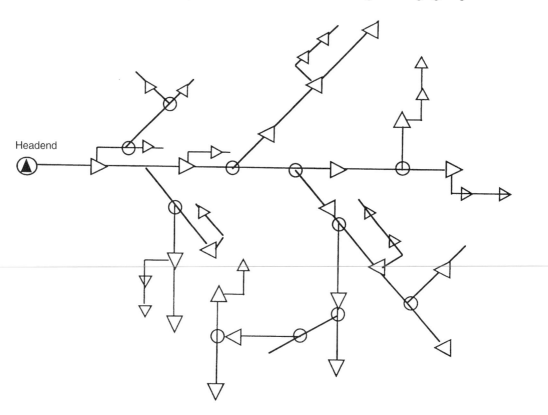

Figure 6.14 Branching Network.

also follows the unity gain concept. Amplifiers following the bridging amplifier are called line extenders. Typically, the trunk is not tapped to serve the subscriber directly but rather is intended to transport signals from one bridger location to the next. Accordingly, the majority of loss on the trunk is cable loss. Conversely, the feeder network contains numerous directional couplers and taps to feed the subscribers. Since the tap and directional coupler losses accumulate faster than pure cable losses, output levels of the bridger and line extenders are significantly higher than trunk amplifiers. Otherwise, many more line extenders would be required, and the cost to build a cable network would escalate. For example, trunk amplifiers are physically placed approximately 2000 feet apart, while line extenders average 900 feet apart.

Since trunk amplifiers, bridging amplifiers, and line extenders are active devices, they require electrical power to function. Currently, a power supply that transforms utility power to 60VAC (60Hz) is used in conjunction with a power inserter that allows placing electrical power on the coaxial cable. Surge-suppression devices are built into the power inserters or the trunk and distribution amplifiers. In addition to these, fusing is also used. As discussed earlier, one method of securing the premium and pay services is to deploy interdiction technology. Under this security scheme, the tap is no longer a passive device but becomes an active component requiring electrical power. Although this technology is very customer friendly, it does dramatically escalate network power requirements.

Areas of concern for management include the proper operation of the active devices, including the appropriate RF power levels into and out of each amplifier; properly operating power supplies; and the physical integrity of the coaxial cables, associated connectors, and passive devices. Deficiencies in any one of these areas degrade signal performance (picture quality) and network reliability. Historical maintenance programs are labor intensive and often require technicians to visit equipment sites to determine that the equipment is working properly. The amount of manpower required under this management scheme is a large drain on the cable operator's technical resources.

To address reliability and picture quality concerns, as they relate to growing trunk amp cascades, the cable television industry began deploying fiber optic technology. The premise is to divide the coaxial network into small serving areas (or service cells) so that the amplifier cascades are held to a low number. The fiber link is composed of three parts; (1) the optical transmitter located in the headend, (2) the optical node that interfaces with the coaxial network near the community to be served, and (3) the optical fiber connecting the optical transmitter and the node. Since each node operates independently of one another, a fault within one cell has no impact on the performance of any other cell. These nodes can be likened to the first trunk amplifier out of the headend in that they are fed directly from the headend, require a power supply and associated power inserter, and contain a bridging amplifier for subscriber distribution. Since much of the microwave transmission links are being replaced by fiber optics technology, these wireless transmission systems will not be covered.

As stated earlier, outside plant has traditionally not been actively monitored or managed. Instead, the customer was the industry's eyes and ears. Today, however, many cable operators are deploying network monitoring and control systems. The first area of significant interest is the fiber optic node, followed closely by power supply monitoring. A third priority is end-of-line monitors that can alert the operator of multiple faults. These are also being deployed as a vehicle for automated annual FCC-required testing.

6.3.3. The Drop

The final portion of the network is the drop, and it includes the cable between the tap and the premise, any passives on the premise, any active device like a house amplifier, and consumer electronics interface devices like the set-top converter. The network design will account for a limited number of outlets on the customer's premise so that the drop incorporates a specified splitter ratio. For those customers requiring more than the usual number of outlets, a house amplifier is required. This unit is typically powered by premise power. Also, with the extended frequencies of modern cable television systems, set-top converters are normally included in the drop. The converter can provide a method for securing premium and pay services as well as provide access to signals that are beyond the tuning range of older television sets. The last item that is part of the drop would be any traps used to secure the more costly services. These are usually placed at the tap for negative traps and at the television set for positive traps. It should be noted that any component of the drop is a potential point-of-failure. Accordingly, proper design and selection of reliable materials are critical to affecting low-cost operations and customer satisfaction. As the more sophisticated services are offered and as customer interface devices become more complex, so method of remote testing and diagnostics of CPE will be required.

6.4. MANAGEMENT OF EXISTING NETWORKS

Current maintenance of CATV networks is divided into two basic categories; demand maintenance and preventive maintenance. Demand maintenance uses simple equipment to identify the source of a problem as a result of a customer complaint. A frequency selective signal meter (SLM) is used to determine which channel the signal is on and the power levels of various signals. Metallic Time Domain Reflectometer (TDR) is used to test coaxial trunks and cable continuity problems, in general. As a general guideline to trouble shooting, if the low frequency signal is Out-Of-Specification (OOS) and the high frequency signal meets the specification, the problem is usually a connector corrosion. On the other hand, if the high frequency signal is OOS and the low frequency signal meets specification, the problem is usually one of continuity. Preventive maintenance is defined as the routine maintenance scheduled at regular intervals to ensure the sanity of the network and minimize customer complaints by detecting potential problem spots and providing necessary correction. The following preventative maintenance activities are the most common among CATV operators:

- Leakage monitoring; performed quarterly.
- Balancing both trunk and distribution amplifiers; performed twice a year.
- Balancing power supplies; performed quarterly or twice a year.
- End-of-line proof of performance; performed monthly.

Cable operators utilize various forms for maintenance activities. These forms, although they are not necessarily industry standard forms, usually carry similar information and may be standard within the same MSO. Among these forms are the following:

- Crew assignment
- Trunk amplifiers card gain
- Line extenders card gain
- Trunk amplifiers record file
- Line extenders record file
- Field discrepancy report
- Monthly vital signs report
- Hub line extender form
- Multichannel distortion report
- End-of-line monitor
- Measured level report
- Engineering video quality form
- System data sheet
- Power supply maintenance record

In addition to the above maintenance activities, most cable operators generate outage procedures and standby policies to ensure smooth operation. These outage procedures outline the planned outages for distribution plant, trunk, fiber and headends, the recommended time for these outages, and the responsibilities of various maintenance and operational personnel during the outages. The procedure also defines parameters and responsibilities related to unplanned outages during business hours as well as after business hours, for distribution outages, minor trunk outages, major trunk outages, and headend outages.

6.5. FUTURE NETWORK CHARACTERISTICS

Recent changes in the dynamics of the telecommunications market, driven by open competition, the information superhighway concept, and deregulation, are having a great impact on cable TV networks. To understand this impact, let us first discuss future services and their characteristics, the challenges these services impose on the CATV industry, and how the industry is coping with these challenges.

6.5.1. Future Services

The future services the CATV industry wants to provide can be divided into three major categories: interactive video, interactive data, and telephony. Interactive video services include video-on-demand, pay-per-view, telemedicine, staggercast, and educational video. Interactive data services include home banking, home shopping, telecommuting, home office, and games. Following are examples of the definition of some of these services as per the Digital Audio Video Council (DAVIC).

6.5.1.1. Movies on Demand (MoD). MoD refers to a network-delivered service that offers the functionality of the home VCR (as a player only) without having to get a copy of the chosen material. The user has the ability to use the following features: select/cancel, start, stop, pause (with or without freeze frame), fast forward, reverse, scan forward, or reverse (both with images), setting and resetting memory markers, showing counters, and jumping to different scenes. Not all of these features are required for the service to be MoD.

Previewing and interactive browsing are typical functions. Data transported to (and presented to) the user can also include information like the user's account. Involved are the end-user (domestic or business setting) and providers of content, service, and network. The content provider and service provider can be different.

6.5.1.2. Near Video on Demand (N-VOD). N-VOD, also known as Enhanced or Advanced Pay Per View, is a specific broadcast application that improves the availability of (typically) movies, without requiring a dedicated point-to-point connection to each viewer.

Basic N-VOD covers pure broadcast of video, in a multiplex manner, with no interactivity between the user and the service/network provider. On a regular satellite or CATV system, all the titles are broadcast all the time, with users merely selecting the channel that provides them with the closest start time, which is eventually supported by an Electronic Program Guide. To effect a "Pause" the user selects another copy of the same title that started at a different time. The minimum pause period is equal to the Stagger Time, which is the difference between two successive start times of the same movie (e.g., 15 minutes).

Intelligent N-VoD (IN-VoD) offers the users a more friendly and effective handling of the Pause feature. Effectively, the IN-VOD application handles the activity of finding the appropriate channel to reselect after a pause is executed. This intelligence may reside in the Set Top Unit (STU) or in the network (e.g., service provider or network provider). There is also some interactivity between the STU and the service or network provider for billing and feature data.

6.5.1.3. Telemedicine. Telemedicine is a service similar to the combination of VoD, multimedia information retrieval application, and video-conferencing application. X-ray images (annotated) may be retrieved by the end-user or distributed by the end-user to other end-users for consultation or further evaluation. In addition, real surgical procedure can be broadcast to students (end-users) or other consultants. These data may be (compressed and) stored by the network or service provider.

6.5.1.4. Home Banking.

Home banking is an application that provides electronic access to offerings available in the typical retail bank, which may include retrieving account balances, making payments to third parties, applying for loans, and browsing through bank offerings. The "end-user" is a banking customer (or will become so) of a bank that is a "content provider." The end-user navigates through certain bank offerings in order to access the offerings.

- Personal banking information (account balances, interest rates, terms).
- Transaction capability (vendor point, direct debit, transfers).
- Other banking services (credit cards, savings).
- Application/request capability (loan application, checks).
- Smart counselor (tailored financial advisor [you spend too much on . . :].

This is a point-to-point bidirectional application, and security will play an important role.

6.5.1.5. Telecommuting.

Telecommuting requires the involvement of end-user (teleworker), service provider, and one collaborator. The user establishes a session via a service provider, activates and controls local and distant application and communicates with a collaborator (through audio, video, and data).

The application provides for the user:

- Directory service, bulletin/message board service
- Conferencing service in real time with two users
- Distribution of information, one-to-one user joint viewing
- Joint editing with control of application
- Signature commutation

6.5.1.6. Teleshopping.

Teleshopping allows the user to browse video catalogues or virtual shops in order to purchase products and services. The user may select items to get more information that may be presented using many different media, for example, video, text, motion video with audio, audio, or graphics (still or animated). After the user has selected a product, he may "order" the product. Once the product has been ordered, the method of delivery depends on the service provider's implementation and user agreements.

The following "players" are involved with teleshopping:

- The end-user uses the application and possibly purchases goods.
- A Value Added Service Provider (VASP) provides this service.

- A constant provider may provide media content during the use of this application.

- A second VASP may provide the back-end of the application.

Table 6.3 illustrates upstream and downstream bandwidth requirements for various interactive data and interactive video services. The expected holding time of each service in both directions is also indicated.

TABLE 6.3   SERVICES BANDWIDTH REQUIREMENTS AND HOLDING TIME

		Downstream		Upstream		
		Bandwidth (bps)	Holding time (min)	Bandwidth (bps)	Holding time (min)	Type
	Navigator	384 K	5	64 K	0.5	B
	Information services	384 K	1	19 K	0.1	D
	Directory	384 K	3	19 K	0.3	D
	Display based marketing	384 K	0.1	64 K	0.1	D
	Home shopping	384 K	7	192 K	0.7	D
	Home banking	64 K	3	64 K	0.3	D
	Multimedia news	384 K	3	64 K	0.3	D
	Government services	384 K	3	19 K	0.3	D
D	Telemetry	19 K	0.1	19 K	0.1	D
A	Electronic billing	19 K	3	19 K	0.3	D
T	Education	384 K	30	384 K	30	D
A	Telecommuting	384 K	60	384 K	60	D
	Online services	768 K	15	128 K	1.5	D
	Magazines, newspapers	384 K	30	10 K	3	D
	Home office	384 K	60	384 K	60	D
	Music on demand	384 K	4	19 K	0.1	D
	Games	384 K	60	128 K	60	D
	Educational games	384 K	60	64 K	60	D
	Info library	384 K	30	64 K	0.1	D
	File transfer on demand	384 K	10	384 K	10	D
	Health info	3,000 K	10	64 K	1	V
	Video (movie) on demand	3,000 K	110	19 K	0.1	V
V	Staggercast (NVOD)	3,000 K	120	19 K	0.1	V
I	Pay per view	3,000 K	120	10 K	0.1	V
D	Educational videos	3,000 K	60	19 K	0.1	V
E	Past TV	3,000 K	30	10 K	0.1	V
O	Home video feed	3,000 K	15	3,000 K	15	V
	Video phone	384 K	10	384 K	10	V
	Non-video phone	64 K	5	64 K	5	T

6.6. IMPLICATIONS FOR THE CATV INDUSTRY

Providing the above new services by CATV poses serious challenges to the whole industry. These challenges could be summarized as follows:

- Demand on bandwidth: More bandwidth is required downstream as well as upstream in order to provide these new services. The bandwidth requirements will depend on the cableco service offering, rate of buy, and service penetration within the served area.

- Two-way communications which will require the activation of the cable system reverse path. If the low split spectrum is used for the reverse direction, the industry has to solve the ingress and egress problem.

- Reliability issues: The CATV network was never perceived as a highly reliable network. Providing these new services, especially telephony, requires very high reliability. The availability objectives for telephony is 99.99 for the access portion of the network (10). This translates to a per line downtime of 53 minutes/year. This objective could not be met with the existing CATV network. Certain upgrades as well as new Intelligent Network (IN) elements are needed to ensure that the network is ready for these new services.

In addition to the above challenges, CATV industry is currently facing new competition from telcos, wireless, and Digital Broadcast Satellite (DBS).

To cope with the above challenges, most of the leading Multiple System Operators (MSOs) are upgrading their networks to position themselves for competition. This starts first with upgrading to 750 MHz systems in order to have enough bandwidth to provide more TV channels. Along with this, the number of fiber kilometers deployed in the network, as well as optoelectronic equipment, have recently increased manyfold to ensure superior quality of signal transmission and delivery as well as higher reliability of the network.

6.7. DEMAND PLACED ON NETWORK MANAGEMENT

As described above, these new services place great demand on network reliability. Telephony is perhaps a good example. Because a person's life and livelihood depend on telephone services, such as enhanced 911 calling, customers and public utility commissions demand a very high level of reliability. Outages must be measured in just seconds per year. In addition, interexchange carriers will demand that providers satisfy tough performance metrics on both reliability and quality. These networks are expected to carry these services with greater than 99.99 percent reliability across traditional franchise borders. As such, building a smoothly functioning multiservices broadband network will take more than upgrading transmission systems and installing a few switches and software programs. Currently, most of MSO fault management is reactive rather than proactive; troubles are usually reported by the customer. MSO cannot afford to continue using the same mode of operation with new services in the new market environment. For MSO to succeed, intense and precise reengineering of operations is required. Proactive network management in a way that will guarantee resolving problems before the switchboard lights with complaints is required. Smart, sophisticated management systems and operational support solutions must be implemented. These OS solutions can no longer be thought of as only a billing system and a headend manager.

6.8. FUTURE NETWORK MANAGEMENT SYSTEMS

Leading MSOs are currently developing and deploying sophisticated network management systems to make use of the vast amounts of information that will be generated by IN elements on their status, activities, and traffic patterns. As such, these network management systems have to monitor and control all network equipment: primary fiber hubs, secondary fiber hubs, trunk amplifiers, power supplies, and coaxial cable TV systems. In the event of fiber cut or optical hardware failure, a backup route or redundant hardware automatically switches in to restore service. The management system monitors all equipment and reports any fault conditions to a central technical action center (TAC). These systems have various security levels in which all users can monitor the network, but only designated technicians have the ability to issue switching or reconfiguration commands. These systems also provide the operator with a comprehensive view of the entire network status at any time. This contributes to more effective problem isolation, resulting in real gains in network operations efficiency and customer satisfaction.

In most cases, cable networks consist of hardware from numerous vendors and are of varying vintage. Since network management issues either have not been addressed or the solution they offer is proprietary and applies only to their particular hardware, integration is a major concern in developing these management systems. It is of great importance that future management systems integrate various stand-alone systems into one so that an operations technician can view the network on one computer screen as a system. The formula for a successful network management system is cohesive interaction between the human operators, the software applications, and the network hardware elements. The system must also be very user friendly, providing automatic updates to operations technicians and allowing them to concentrate on solving problems rather than operating the network management system.

6.9. OTHER CATV MANAGEMENT SYSTEMS

Figure 6.15 illustrates the integration of various existing databases into a management system. The integrated system draws on various databases, including customer information, billing and accounts receivable, service and work orders processing, and system information databases. The system then generates reports to help control inventory, to assist technical staff, and to manage addressable systems and general system reports.

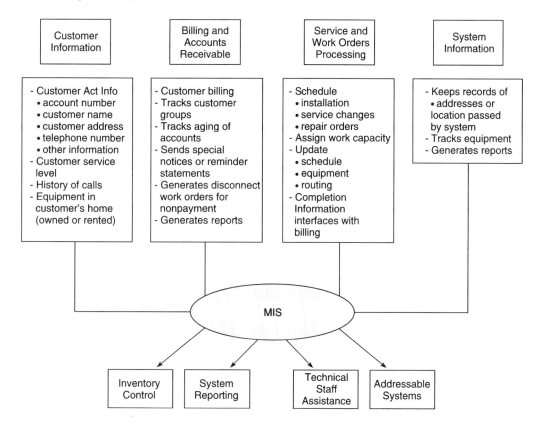

Figure 6.15 Management Information System.

6.10. OPERATIONS SYSTEMS (OSs)

OSs and business support systems (BSS) are two telco industry buzzwords used to refer to the infrastructure that has been put in place to monitor networks, provision new services, provide customer support, and bill for services rendered. The following is a set of standards developed by ITU-T (formerly CCITT) to describe various functions to be performed by OSs:

- *Configuration management*—The ability to troubleshoot the network remotely, without having to send service personnel just to find out which network component has failed.
- *Fault management*—Similar to cable TV status monitoring, it allows the user to perform remote diagnostics.
- *Performance management*—In order to anticipate problems before they actually occur, the network is able to collect statistics and develop trend lines related to how the network is performing.

- *Accounting management*—The ability to create accurate billing statements for the amount of time the network was used.
- *Security management*—Only authorized users are allowed to access the network.

Figure 6.16 illustrates the latest MSO's thinking in the area of management and operation support systems. It calls for linking various systems and databases into one system, which can help produce computer-aided dispatch systems, global positioning systems, and automated work management systems. The following example examines how such integrated OSs can enhance provisioning and maintenance of new services in a complex, multiservices network.

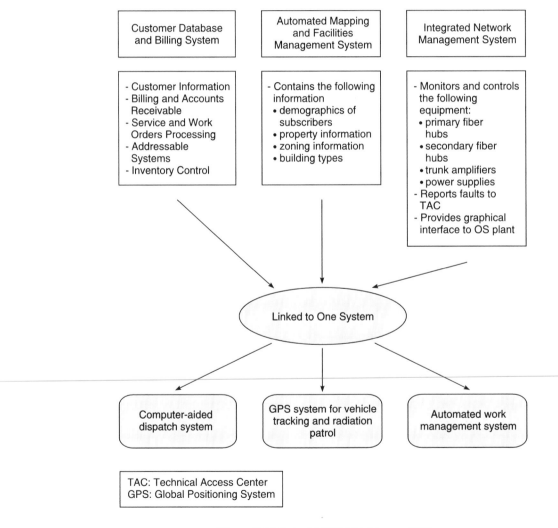

Figure 6.16 Latest MSO Thinking re OSSs.

Consider the provisioning of telephony service. Today it is usually done in discrete steps. The customer services representative (CSR) takes a customer's order and ends the call. Then the order is sent to an order processing system that divides the order into subtasks; the subtasks are assigned to different personnel or downstream systems to identify available facilities or install new ones, turn on service, set up the billing account and so on.

Only after the subtasks are underway is it possible to accurately determine when service will be activated. If any service order information is missing or incorrect, the order is delayed pending manual intervention.

If, however, intelligent broadband network elements and the information they generate are used effectively, most tasks can be accomplished during the customer's call. As the CSR electronically enters a service order, the OS can determine the facilities needed, communicate directly with the IN elements to determine the availability of resources and facilities, issue the proper electronic provisioning commands, and set up a billing account. If the appropriate resources are available, the CSR then tells the customer the time that service will begin.

References

[1] Schwarz, Richard, "Broadband Networks Demand Operations Support," CED, June 1995.
[2] Bartlett, Eugene R., *Cable Television Technology and Operations: HDTV and NTSC Systems*, McGraw-Hill, New York, 1990.
[3] Sucharczuk, Guy, "Network Management Systems Advantages and Implementations," 1994 NCTA Technical Papers.
[4] Stanzione, Dan, "Operations Systems: A perspective," *AT&T Technical Journal*, July/August 1994.
[5] Brown, Roger, "Cable Struggles to Build an OSS," CED, July 1994.
[6] Means, Ed, "Tomorrow's OSS Is Here Today," *Telephony*, March 1995.
[7] Wright, Steven, "Operational Support Systems," CED, March 1995.
[8] Millitzer, Thomas, "MIS for CATV Systems: An Important Part of Overall Operations," *International Cable*, July 1993.
[9] Stone, David, and Clifford, Courtenay, "Managing Operations in a Video World," *TE&M*, July 1993.
[10] Bellcore TA-NWT-000909 Issue 2, December 1993.

Mike Ahrens
Bellcore Software Systems

Chapter 7

Architectural Integrity as a Countermeasure to Complexity in the Telecommunications Management System Life Cycle

Over the past decade, several significant strides have been made toward improving key aspects of the TMS (telecommunications management system) life-cycle process. However, we must still ask ourselves if we have fundamentally solved the *underlying problems* in removing the lag involved in bringing TMS solutions to market concurrently with the rollout of a new Service and Network Initiative (S&NI). Most of the efforts over the last five years or so to improve the art of TMS design and development have tackled some of the "accidental" complexity, but have not addressed head-on the "essential" complexity in TMS solution architecture. The key to reducing this lag is to *systematically identify, and manage, the fundamental sources of complexity in the initial stages of the TMS life cycle for a new service and network initiative.* Because of the inherent complexity in defining complete TMS solutions, an architecture team is required. We will examine the challenges faced by the TMS solution architecture team and address the life-cycle process as a learning cycle. Only by understanding the structure and dynamics of the organizations involved in the life-cycle process, and then attacking the *essence* of the complexity in the flow of information among organizations within the overall S&NI life cycle, can the time-to-market for TMSs truly be improved.

7.1. INTRODUCTION

A central issue currently facing Telecommunications Management (TM) practitioners is how to anticipate and respond to business, market, technology, and organizational forces faster and more effectively than the competition. Since the late 1980s, network providers have recognized that Telecommunications Management System (TMS) solutions represent a bottleneck in the process of realizing new business opportunities. When a network provider is carrying out strategic business initiatives—bringing to market new services based on new and emerging network technologies—TMS solutions are known to lag the availability of the elemental network/service infrastructure.

7.1.1. The Quest for a Breakthrough

Over the last decade, several significant strides have been made toward improving key aspects of the TMS life-cycle process. These developments have occurred in several notable areas, including *layered management modeling, logical data modeling, object-oriented modeling, standard interface protocols, distributed computing platforms*, and *graphical user interfaces (GUIs)*. Improvements in these areas, and in others, have clearly improved the overall practice of providing network management solutions. Each of these efforts taken individually, and certainly the sum total of all of these efforts taken collectively, deserve ample acknowledgment. However, we must still ask ourselves if we have fundamentally solved the *underlying problems* in bringing TMS solutions to market concurrently with new network technology and service initiatives. Moreover, have we really been effective in articulating a definitive statement of these *underlying problems*?

Many of the efforts just mentioned have been successful in adding a level of discipline to the process of building TMS solutions. However, in most cases they represent only modest improvements in dealing with the fundamental, underlying challenges in the TMS life cycle. This is especially true in contrast to the rate at which breakthroughs have occurred in other areas of telecommunications. Particularly in the area of telecommunications network architectures, breakthroughs have been both numerous and extensive. Telecommunications network designers and engineers have been able to highly leverage order-of-magnitude breakthroughs in semiconductors, microelectronics, optoelectronics, data compression, and related standards. This contrast is especially important given that many decision makers in telecommunications companies come from more of a "hardware" background. These key stakeholders tend to grow impatient with the relative lack of progress in improving the software system life cycle.

Unfortunately, by comparison, TM planners are faced with the prospect of no "silver bullets"[1]—no single or series of developments, in technology, engineering, or management technique that can do for software what various breakthroughs in the last few decades have done for hardware. It is frustrating to admit that software in general, and TM software in particular, remains, by its very nature, riddled with fundamental problems. Even if we look back over the last several decades, can we

---

[1] Frederick P. Brooks, Jr., "No Silver Bullet; Essence and Accidents of Software Engineering," *Computer*, April 1987.

really claim that much fundamental progress has been made in reducing the high degree of complexity inherent in defining and building TMSs?

7.1.2. Accidental versus Essential Complexity

Over two decades ago Fred Brooks "wrote the book" on complex software engineering.[2] In a later work[3] he explains why the quest for a *silver bullet* (a fundamental breakthrough in productivity and effectiveness) is so elusive for software systems. Brooks differentiates between *essential* and *accidental* complexity. In these terms, most of the above-mentioned efforts to improve the art of TMS design and development have tackled some of the "accidental" complexity but have not addressed head-on the "essential" complexity in TMS solution architecture.

Essential complexity consists of difficulties inherent in the nature of software, while accidental complexity pertains to difficulties that are manifest in the *current* state of the practice but are not inherent. Brooks explains that while progress can and is made over time as a result of attempts to remove accidental complexity, essential complexity is never removed. Essential complexity can instead only be *coped with* through "persistent, unremitting care" and through sustained application of project *discipline.* The concept of accidental versus essential complexity is quite useful to our understanding of the practice of building TMSs.

What then are the sources, or causes, of the complexity inherent in the end-to-end life-cycle process for defining, building, and deploying TMSs? Some of this complexity is due to the very nature of software systems. Since TMSs are, at their core, software, we will briefly look at some key aspects of complexity common to all software systems. Other aspects of TMS complexity have more to do with how the TMS life cycle fits into a broader business context. This broader context will be referred to as the Service and Network Initiative (S&NI) life cycle—that is, the overall life cycle for bringing new service offerings, carried on new network technology, to market. The relationship between the TMS life cycle and the S&NI life cycle is shown in Figure 7.1.

S&NI Life Cycle
(e.g., Residential BB, ATM CRS, Wireless Local Loop)

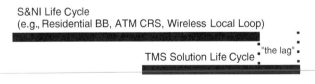

TMS Solution Life Cycle "the lag"

Figure 7.1 Relationship between TMS and S&NI Life Cycles.

For clarification, it is important to note that we are not focusing on the issue of delivering a single, stand-alone TMS application, such as an element management system for managing the HFC (hybrid fiber coax) equipment in a residential broadband loop, or a vertically integrated element/network management system for ATM

[2] Frederick P. Brooks, Jr., *The Mythical Man-Month; Essays on Software Engineering*, Addison-Wesley, 1975.
[3] "No Silver Bullet."

(asynchronous transfer mode) CRS (cell relay service). Instead, we are referring to a complete solution, adequate to support, for example, the rollout of a residential Internet access service offering across an entire metropolitan area. Such a solution likely consists of five to nine major TMS applications that are networked together into a complete TMS solution architecture. To successfully support the commercial rollout of a major S&NI project, such a solution architecture typically supports automated flowthrough not only for service activation, but also for network creation and service assurance, including automation of network inventory management, autodiscovery of equipment inventory, alarm surveillance, and trouble management applications.

Our overarching objective is to understand what is required to reduce "the lag" between (a) when the network technology and service offerings would otherwise be ready (apart from TMS support); and (b) when it is realistically possible to roll out the network technology and service offerings in a cost-effective manner, complete with commercial-grade TMS solution support. The central thesis of this chapter is that *the key to reducing this lag is to systematically identify, and attempt to manage, the fundamental sources of complexity in the initial stages of the TMS life cycle for a new service and network initiative.*

7.1.3. Complexity Inherent in Software Systems

Since coping with the complexity essential to software systems is central to any attempt at streamlining the production of TMSs, we will briefly summarize Brooks' description of four basic factors that make software production inherently formidable. These factors are *conformity, malleability, invisibility, and intricacy.*[4]

Software is forced into *conformity* with various types of boundary conditions in seemingly arbitrary and capricious ways by the many human institutions and systems to which it must interface. Organizations bring differing assumptions and expectations, and they force the systems they make or buy to fit in the specific context of the organizations and corporate cultures they represent. For TMSs, there is tremendous pressure to "pave old cow paths," that is, to apply a method of raw mechanization whereby existing manual processes are simply codified into programmed processes within the TMSs. In the context of the S&NI life cycle, this is especially problematic if opportunities (availed by the new network technology or the new service offerings) to simplify existing processes are not recognized and exploited. However, conformity to existing business practices is difficult to overcome unless a focused, concerted effort is made to overcome this natural tendency.

Malleability addresses the fact that since software systems are relatively amenable to change in many dimensions, they are constantly subject to pressures for change. The pressure for change comes primarily in two forms: pressure to *expand* its utility beyond its original domain of application, and pressure to *extend* its useful life beyond that of the computer platform for which it was originally written.

---

[4] Brooks uses the term *complexity*; since we are using this term somewhat more broadly, we use the term *intricacy* in place of Brooks' more specific *complexity*.

Hardware systems, by contrast, are typically expected to permit change in well-defined dimensions (e.g., memory expansion boards, peripheral adapter ports, hardware expansion slots). Furthermore, in the hardware equipment realm, it is much more clearcut to determine when a piece of equipment has reached capacity exhaust, or when the equipment has been fully depreciated, or cannot be cost-effectively upgraded to support a new type of functionality.

Regarding the *invisibility* of software design, Brooks characterizes software as invisible and unvisualizable. He cites the lack of tools that truly permit the human mind to use some of its most powerful conceptual tools. This invisibility especially cripples the ability of multiple designers to communicate with each other and with "downstream" participants in the software production life cycle. For TMSs, individual piece-part applications may not be so difficult to visualize, but understanding the required interplay and possible efficiencies and synergies across an architecture of TMS applications is often quite elusive.

Brooks describes many aspects of software *intricacy*. Software is composed mainly of nonrepeating parts. The interactions between the individual parts can be characterized by numerous states, and the interactions among parts tend to be nonlinear. A result of the nonlinearity is that scaling up a software system involves an increase in the number and type of different parts, as well as interactions. For TMS solutions, it is often difficult to recognize potential overlaps in related parts across multiple TMSs. TMS solutions often involve a threaded flow of information through many related parts. For example, a recently implemented autodiscovery solution for residential broadband network equipment requires coordination across the equipment states tracked in as many as five interrelated TMSs.

On an organizational level, intricacy leads to communications problems among team members, due in part to difficulty in comprehending the complete behavior of a total system solution. Perhaps the most perilous result of intricacy on a large project is that it limits the extent to which the *conceptual integrity* of a system design can be preserved throughout the project life cycle. Brooks put it this way: "*I will contend that conceptual integrity is the most important consideration in system design.*"[5] Conceptual integrity is the essence of architecture. It is what takes individual component pieces and integrates them together to create a meaningful whole that provides a solution that is not possible from the collection of components alone. Unless it is managed systematically, intricacy is like *entropy*—it causes a system to degenerate to a state characterized by maximum disorder and minimum energy.

7.1.4. Conway's Law: Linking Organizational to System Architecture

Although terms like *software engineering* and *systems engineering* are often used to describe the work of practitioners in the software system life cycle, many of the phrases in the above paragraphs sound more appropriate to the field of organiza-

---

[5] *The Mythical Man-Month*, p. 42.

tional sociology. While this may be unsettling to some, it is important to appreciate the connection between TMS solution architecture and organizational structure and dynamics. As we will elaborate upon more fully in section 7.3.1, the primary underlying factor that contributes to the complexity of complex systems is the requirement for focused and sustained coordination between large numbers of experts across various disciplines. Thus, reducing and managing essential complexity in complex systems requires that organizational structure and dynamics be managed *as an integral part of* the architecture of the system.

Perhaps the best way to sum up the connection between software systems, and organizational structure and dynamics is Conway's Law: "*Organizations that design systems are constrained to produce designs that are copies of the communications structures of these organizations.*"

A key to streamlining the process of rolling out integrated TMSs in a timely manner is to strategically evaluate the complete organizational context in which the TMSs will live. This includes the organizational structure and dynamics of all stakeholders in the S&NI (Service and Network Initiative) life cycle as a macrocosm. Only by understanding the structure and dynamics of the existing organizations, and then determining how to improve the flow of information, or moreover the shared corporate *learning*, among key interacting organizational components, can significant progress be made. In other words, only by attacking the *essence* of the complexity in the flow of information among organizations within the overall S&NI life cycle can the time-to-market for TMSs truly be improved.

As alluded to previously, many of the network and service initiatives invite fundamental changes in the way an organization is structured in terms of business processes. Properly architected TMS solutions are an integral part of realizing these opportunities. Yet, existing TMS solutions are in place and are intricately interwoven into the organizational fabric of the organization. New TMS solutions that leverage opportunities availed by new network technology are effective only to the extent that they can be integrated into the evolving organizational framework within which they will exist. This is particularly challenging in that the organizational structure and dynamics will typically be undergoing a transformation due to rapidly changing market pressures.

First, we will look at the S&NI life cycle so that we can understand the context in which TMS solutions must be architected. With this as a backdrop, we will then look at models of how learning needs to take place for TMSs to emerge out of this life cycle at the right time, with an architectural design that provides a complete solution to the appropriately defined problem.

7.2. TMSs IN THE CONTEXT OF THE S&NI LIFE CYCLE

Though the term *operational support systems* is not used as widely as it once was, this colloquial alias for TMSs is a sober reminder to TMS practitioners that we produce systems whose primary purpose is to *support* other aspects of a network provider's business. Many are quick to remind us that TMSs are but the "tail on the dog." To better characterize the "tail," then, let us first look at the rest of the dog!

Figure 7.2 provides a simplified, generic view of the S&NI life cycle for a Network Provider (NP). The process is more complex than is portrayed in the figure, but the intent is to capture some key aspects of the life cycle, particularly those that bear most directly on the TMS life-cycle process. This representation of a S&NI life-cycle process is not intended to exactly match the process followed by any specific company but is rather an attempt to portray a general example. The specific process followed by any particular network provider (and their selected network equipment and TMS suppliers) may differ in particular details from this generic S&NI life-cycle process. However, any differences are minor compared to the aspects that all network providers will have in common when executing the S&NI life-cycle process.

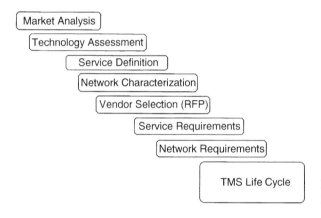

Figure 7.2 NP Service and Network Initiative (S&NI) Life Cycle.

While, in concept, such a life-cycle process has been employed in the past, it seems that only recently have attempts been made to execute a complete instance of this cycle in a narrow, focused time frame (e.g., 18–24 months). As competitive pressures continue to squeeze the cycle time for this process, the NPs will likely strive to define this process more formally out of necessity for success, if not survival.

7.2.1. Market Analysis

A typical S&NI life cycle begins with *market analysis* to determine fundamental issues such as whether an NP can compete in a new or emerging market, how the market is segmented, who are the competition, and so on. At this stage, TM issues may be considered in terms of how they will likely play into cost versus quality-of-service tradeoffs. If TM issues are addressed at all at this stage, they tend to be addressed in general terms, relative to the overall business functions and needs required. Typically, at this stage, TM issues are addressed in regard to what existing competitors offer, or what customer expectations may be, for example, customer management of data services.

7.2.2. Technology Assessment

New technologies are often an enabler for market opportunities. Keeping abreast of rapidly evolving and maturing technologies is vital to an accurate assessment of key NP business opportunities. As such, timely *technology assessment* must follow close on the heels of an initial market analysis. It is likely that these two stages in the life cycle are far more intimately intertwined than Figure 7.2 would indicate. Typically, however, technology assessment focuses on the network technologies, with little attention given to TM issues. This is often warranted since basic issues of bringing the technology from a lab prototype to a field-ready commercial system tend to swamp TM issues. However, this is often where some of the later problems begin to creep in.

Analogous to key learnings from the manufacturing industry, a concept akin to "design for maintainability" is beginning to demonstrate relevance in the design of telecom network equipment. However, NP "equipment engineers" still exhibit a strong tendency to give limited attention to "OA&M" (Operations, Administration, and Maintenance) features in equipment—certainly at the earlier stages in the S&NI life cycle.

It is at this stage that an NP can begin to factor the TM solution infrastructure into the overall cost equation. Certain technology features may potentially drive operations costs out of the business, for example, dynamic TSI (time slot interchange) in NGDLC (next-generation digital loop carrier). However, these potential cost reductions can be realized only if the corresponding investment is made in a TMS infrastructure, for example, a management system capable of assigning CRVs (call reference values) and managing load balancing across multiple VRTs (virtual remote terminals), and in TM features of the equipment. (Reduced costs for dispatch due to accurate, remote testing can be achieved only if the NEs are equipped with the appropriate drop test unit.)

7.2.3. Service Definition

Service definition is principally an attempt to narrow the scope of potential offerings that will meet the perceived market opportunity. What can the NP offer cost effectively that provides a means of differentiating the NP from its competition, while meeting a customer need and producing revenue? At this level, TM capabilities may be looked at in more detail. However, a really detailed and meaningful scrutinization of potential TM functions to support service offerings typically must be deferred to the next stage. This is because the services defined at this stage can potentially be carried on a variety of network technologies, and a detailed assessment of potential TM functions requires fairly detailed knowledge of a specific network technology. In the future, an improved process would factor TM issues more explicitly into all of the early stages of the S&NI life cycle, but particularly into the service definition stage. For example, a complete service definition, from a TM point of view, should include some basic requirements for service-level testing, even if independent of network technology.

7.2.4. Network Characterization

If an NP chooses to make use of an RFI (Request for Information) in more formally evaluating network technology, this will occur at the *network characterization* stage. Whether or not an RFI is pursued, this stage defines the capabilities of one or more network technology solutions that can be assembled to provide the defined services. This is where tradeoffs between competing technologies are evaluated in detail. Often this is done on the basis of qualitative and quantitative assessments of the equipment suppliers' ability to meet specified cost points for a representative deployment cross section. The amount of technical detail analyzed in this stage increases dramatically over earlier stages.

As will be developed more fully in subsequent text, if the overall S&NI life cycle is to be streamlined, a key requirement will be to draw expertise from the later stages of the life cyle into the design, analysis, and decision making of the earlier stages. The network characterization stage offers the most pronounced opportunity for this to occur.

Traditionally, this stage is dominated by network equipment engineers. Although many of them have some appreciation for TM issues, for various reasons they typically do not draw in TM expertise at this stage. In fact, they seem to prefer a clean handoff ("throw it over the wall") of their engineering design to the downstream stages of the life cycle. To put it in another light, there has traditionally been a strong desire to "buffer" the network equipment engineers from getting dragged into the "messy" details of the "OS" world. However, there is an increasing awareness that equipment engineers need to share with the TM planners at an earlier stage more of their detailed knowledge of how the equipment will operate. At the same time, TM planners need to selectively raise the awareness of the equipment engineers regarding tradeoffs in potential TM features within the equipment, and how these tradeoffs will impact equipment operation, as well as overall service availability as perceived by the end customer.

Based on lessons learned in the manufacturing sector, the benefits of designing for operability/maintainability have demonstrated to engineers clear advantages in inviting input "from the shop floor" (i.e., from those further downstream in the life cycle, for example, in development and testing). In section 7.3.5, we will look at opportunities for the S&NI life cycle that parallel this and other "lessons" drawn from the manufacturing sector. In particular, we will examine the waterfall model—a common representation for the traditional life-cycle process involving information handoffs between upstream and downstream process stages. Our specific focus will be on how a traditional waterfall process can be augmented to increase the degree of information sharing at earlier stages in the process. The benefits of improving the handoffs between stages include reducing the overall cycle time and increasing the accuracy of shared information at earlier stages, where design changes are significantly less costly to make.

7.2.5. Network Equipment Supplier Selection

A major reason why it is so important to give full treatment to TM issues in the network characterization stage is that the *supplier selection* stage follows immediately on the results of network characterization. Moreover, network characterization produces the high-level (compared to other stages) requirements that form the basis for the RFP (request for proposal) on which vendor selection depends. Even if NP requirements in the RFP include TM requirements, for example, at the NEL (Network Element Layer), and EML (Element Management Layer), it often proves difficult to sustain adequate pressure on the supplier throughout the S&NI life cycle to realize timely procurement of TM features in the supplier equipment. However, if the NP RFP lacks sufficient detail on TM requirements, then it is virtually certain that TM features will be relegated to later releases (if provided at all). This in turn has a ripple effect all of the way down the chain of the S&NI life cycle, deep into the TMS life cycle.

The flip side of this discussion is that if the NP RFP contains TM requirements that are overly optimistic, excess energy and resources can be expended at the wrong time by all involved parties. If the right set of TM expertise is involved in the network characterization stage, then the likelihood of the RFP requirements being on target will be far greater. Involving the right expertise means "reaching ahead" to tap the knowledge of experts who are traditionally involved only in much later stages.

7.2.6. Service Requirements

Once a supplier is selected, detailed requirements need to be developed in conjunction with the supplier and the NP marketing organization to accurately specify *service requirements*. These service requirements will drive key network requirements and will tend to dramatically impact cost points for the entire duration of the S&NI life cycle. At this stage, since many of the details of the network technology are known, it is possible to make definitive tradeoffs as to whether specific TM features will "pay for themselves" or negatively impact cost points. As NPs coming from a tradition of "universal service," including rich TM functionality and high TMS availability, contemplate competition with other providers having a much lower TM profile, factoring TM issues into the detailed service requirements becomes a key business issue. Will a higher quality portfolio of TM functionality sufficiently differentiate an NP so that the additional cost associated with providing the richer TM features can be justified?

At the same time that the network technologies and services are rapidly changing, the basic techniques needed to manage these networks and services are also changing. Furthermore, the cost structure associated with performing certain management functions is being modified, driven by changing attributes of the network technology, changing cost elements in the labor market, changing service quality perceptions by the end-user, and competitive forces. These issues must be factored into decisions about what network management features are needed in the NEs, in the EML, the NML (Network Management Layer), and so on. Two specific examples are:

1. The role of loop-back testing, for both telephony and interactive services, on HFC (hybrid fiber coax) or FTTC (fiber to the curb) networks.
2. Traffic data collection for GR-909 technology, for both telephony (with concentration on loop access facilities—dynamic TSI) and interactive services.

Key questions that must be raised at this stage include:

1. How do TM costs weigh against service revenue projections?
2. How do the benefits of providing the TM capabilities factor into the relative value perceived by the customer and the customer's willingness to pay?
3. Are the TM capabilities necessary to remain competitive?

A key challenge in this area is the need to understand the network technology sufficiently, and its potential TM features, in order to determine what TM features will adequately maintain the network (and the services it supports). At the same time, sufficient data must be gathered to support the costs associated with providing these TM features. With a new technology, this can prove nearly impossible since the data needed to generate an adequate business case (e.g., how the proposed TM feature will save operational costs) likely does not exist. The only hope for arriving at meaningful data is to rely on TM expertise (from the tail end of the S&NI life cycle) with existing technologies, and, based on a comprehensive knowledge of the new technology (obtained from very close interaction with NP and supplier engineering expertise), to extrapolate data for the new technology.

7.2.7. Network Requirements

Until fairly recently, it would have been more accurate to draw the *network requirements* stage as two distinct stages: NE requirements (i.e., basic telecommunications switching and transport requirements) and NE telecommunications management (TM) requirements. While the two sets of requirements can still be distinguished, they are shown as one integrated stage because they are increasingly being packaged and delivered as a single document. One reason why they have been separated in the past is that defining NE TM requirements typically depends on the NE telecommunications requirements being at a relatively mature level. That is, the NE TM requirements almost by necessity lag the NE telecommunications requirements. However, the current state of practice is that NE TM requirements are closely coordinated with NE telecommunications requirements, at least for critical NE TM features.

As a general rule, the most critical TM functions in the NEs tend to be alarm surveillance and remote activation of provisioning actions and parameters. These functions are the most mature in terms of standardization, and they tend to be worked earlier in the life cycle for a new technology. Even for newer technologies, for example, ADSL (asymmetrical digital subscriber line) and SDV (switch digital video), these functions tend to get the earliest and most in-depth attention. Other functions are increasing in importance but are still somewhat less established, especially for newer technologies. These functions include autonomous inventory reporting, performance monitoring, software download, memory restoration, and traffic data collection.

Just as defining TM requirements for the NEs depends on a certain level of maturity being reached with the basic network technology requirements, driving to detailed TMS requirements depends on a level of maturity of the NE TM requirements. Even so, the forward progress of NE telecommunications technology, and NE TM requirements can be thought of as the "push" that drives TMS requirements forward. After all, a key aspect of improving the S&NI life cycle is making the TMS solutions available as close as possible to when the technology is deployed.

If NE technology availability is the major force driving the TMS life cycle, additional forces must be taken into account. These forces include the requirements imposed on the TMS life cycle by the personnel responsible for ongoing TM (operations) process and TM (operations) centers. These forces have sometimes been referred to as the "tyranny of the SMEs." This somewhat exaggerated term refers to a tendency for TMS users to control the budget for TMS enhancements, preferring new display screens, or new fields on existing display screens, or new keystroke-saving enhancements to more fundamental modifications required to support new network and service initiatives.

A major challenge in defining TMS requirements for management of a new network technology is that the "push" to trial and deploy the new network technology must not get too far out in front of the "pull" of the TM process and center requirements. Explicit effort must be made to ensure that the "push" and the "pull" are complementary rather than working at cross purposes. These forces can be at odds because the team responsible for the S&NI life-cycle process may not be properly integrated with the various teams that are chartered with "reengineering" the TM processes and centers. The TMS SMEs are typically not centrally involved in the S&NI process at the early stages and do not have it as a high priority. They tend to be driven by other, more immediate needs. Integrating the skills and knowledge of the members of the S&NI team with those of the various teams responsible for evolving TM processes and centers is a key factor in decreasing the overall S&NI cycle time.

7.2.8. Key Learnings for Introducing TMSs in the S&NI Life Cycle

Much more can be said about the relationship between the stages in the S&NI life cycle "upstream" of the TMS life cycle, as well as the stages downstream, in and beyond the TMS life cycle. We will address some of these issues more fully in sections 7.5 and 7.6. At this stage, however, we offer some insights on the art of successfully introducing new TMS systems. These "lessons" are drawn from:[6]

---

[6] Eberhardt Rechtin, *Systems Architecting; Creating & Building Complex Systems*, Prentice-Hall, 1991, p. 120.

- In introducing technological and social changes, *how* you do it is often more important than *what* you do.
- Recent studies have found that the vast majority of new technology introductions are unsuccessful, principally due to a failure to understand the impact on the human infrastructure.
- If social cooperation is required [which is *always* the case for TMS system introduction], the way in which a system is implemented and introduced must be an integral part of its architecture.

To which we offer the following two statements by way of corollary:

- A great technology solution that is poorly introduced and integrated into the culture is *much worse* than an average technology solution that is well introduced and integrated.
- To successfully introduce significant changes to a TMS architecture, significant changes in organizational structure and dynamics must occur in parallel.

Moreover, experience from other industries points to the need for a learning process specifically tailored to preparing an organization for introducing new technology solutions and the required, accompanying changes to organizational structure and dynamics:[7]

> Our research has taught us that there are many things that ought to be done first to prepare one's organization for [these] new technologies. Managers ought to begin by attempting to reduce confusion, and by motivating and managing the learning process in their organization.

A family of TMSs designed or enhanced to accommodate a new technology (e.g., broadband, PCS) must be accompanied by a corresponding, complementary "system" of business processes and preparedness throughout all of the "downstream" recipient organizations that will use or in any way be associated with the TMSs. Once again, this phenomenon has been recognized in other industries:[8]

> To manufacture and produce a major complex system requires a comparably complex system—a production complex—of interconnected supply, manufacturing, assembly, test, and distribution facilities; information systems for monitoring and controlling the processes; and skilled operators and managers to assure its efficiency, quality, and responsiveness to the projects it supports.

---

[7] Robert H. Hayes, Steven C. Wheelwright, and Kim B. Clark, *Dynamic Manufacturing, Creating the Learning Organization*, Free Press, 1988, p. 270.

[8] *Systems Architecting*, p. 133.

By analogy, our "complex system" is the suite of required TMSs, and the "production complex" is all of the "machinery" within a network provider needed to receive, implement, administer, and utilize the TMSs on a going forward basis. The job of architecting TMSs does not end with the delivery of a high-level specification to developers or even the delivery of a set of tested systems to the users. To be effective, it must include a complete set of business processes for integrating the TMSs into the changing and evolving network provider business.

7.3. ARCHITECTURAL INTEGRITY AS A COUNTERMEASURE TO COMPLEXITY

The driving impetus for improving the TMS life cycle is the need to have TMS solutions available "just-in-time" to support the rollout schedule S&NIs. Each S&NI is a complex business undertaking, with many dimensions evolving simultaneously. To be successful, the total TMS solution must be defined to be resilient in the face of key forces including:

- *Market Forces*—what markets (customers, services, distribution channels) the NP wishes to serve relative to the competition.

- *Business Forces*—what relationships the NP has with other carriers (subsidiaries, partners, competitors).

- *Network Technologies*—what core infrastructure the NP chooses in which to make major capital investment, as the vehicle for carrying the primary customer services.

- *Organizational Structure and Dynamics*—how the NP structures its internal organizations and orchestrates their interactions to carry out its key business processes in a cost-effective, streamlined, competitive manner.

- *TMS technology*—the hardware, middleware, communications, and computing environment the NP chooses on which to build and operate its TMSs.

If only one or two of these forces were changing simultaneously, then the situation would be considered challenging. However, the present climate involves major shifts in each of these areas. We are in the midst of a sea change in the telecommunications industry. We have entered what will likely be a sustained period of time when several or all of these forces will be simultaneously active in the TMS life-cycle process. The resulting problems that a complete TMS solution must address are so far-reaching and encompassing, and the interdependencies between various components of the problem domain are so intricately complex, that traditional approaches to defining a complete solution to the problem tend to break down.

It is not that the tried-and-true problem-solving (or solution-creating) approaches are wrong or inadequate in and of themselves. Instead, we need to realize that the approaches we have come to depend on are proven in a classic domain of defining relatively noncomplex (single) system solutions. However, they are not adequate to envelop the complexity of the required "meta-system," or "system-of-systems" solutions. Network providers require solutions to large-scale, highly complicated problems when they attempt to bring full TMS solutions to market "just-in-time" to support the rollout of new service and network technology initiatives.

Providing complex, integrated solutions of this nature requires a different approach to structuring the TMS life-cycle process. Successfully defining and building solutions to complex problems requires an overarching architecture approach that is more than just joining a linear ensemble of individual piece-part solution components. Attempting to assemble a solution to a complex problem without the requisite *architectural integrity* results in a mere collection of unharmonized components. Even worse, when an attempt is made to create a solution for a truly complex problem, complexity escalates during the solution-creation process, due to unanticipated, dissipative interactions among components. If the energy required to create the total solution is not properly focused and managed throughout the life-cycle process, energy instead tends to get spent on addressing (but never truly resolving) internal conflicts arising among component pieces that do not have clearly defined and articulated interrelationships.

7.3.1. Characteristics of Complex Systems

A thing is complex when it exceeds the capacity of a single individual to understand it sufficiently to exercise effective control—regardless of the resources placed at [his] disposal.[9]

A system can be defined as "a set of different elements so connected or related as to perform a unique function not performable by the elements alone."[10] If a system has a single person fulfilling the role of system architect, then it can be classified as a *simple* system. However, when the boundaries of the problem space are enlarged, and the System (i.e., meta-system or system-of-systems) providing the solution is composed of more than one (simple) system, then it becomes increasingly difficult for a single person to fulfill the role of architect. Such a meta-system, or System, is considered a *complex* system.

Complex systems are readily recognized by some key characteristics.[11] They consist of many parts, not only in number, but also in type. In other words, the many parts are mostly nonrepeating. The relationships between the parts are not just straightforward one-to-one interactions, but are rather one-to-many, many-to-one, and many-to-many relationships. There are nested relationships involving primary, secondary, and tertiary interactions both upstream and downstream of particular components. A useful metaphor is a plate of spaghetti: you can't pull out a single meatball or a single strand of pasta without getting more than a fork full!

[9] William L. Livingstone, *The New Plague: Organizations in Complexity*, F.E.S. Limited Publishing, 1986, pp. 1–11.
[10] *Systems Architecting*, p. 7.
[11] *The New Plague*, pp. 1–13.

Complex systems involve many disciplines, drawing on and creating a need for integration across areas of expertise that are conventionally quite separate. For example, management of SONET transport networks requires an integrated expertise drawn from what have traditionally been separate "inside plant" and "outside plant" experts. In a related vein, complex systems also incorporate knowledge from various technologies and draw upon information about diverse fields of science.

The emerging FTTC-based telephony and digital video access/distribution architectures are a good example on this point, since they draw on many evolving technologies, including MPEG (Motion Picture Experts Group) 2, ATM, SONET, Q.2931, TR-909/303, optical splitting and amplification, power distribution, CATV (Community Antenna TeleVision) devices (splitters, amplifiers, taps/combiners), video modulation techniques such as QAM (Quadrature Amplitude Modulation) and CAP (Carrierless Amplitude Phasing), and advanced distribution cable technology.

To implement complex systems typically requires coordination across many organizations within a company, many companies within an industry sector, many sectors within an industry, and sometimes many industries. Moreover, the organizations, companies, sectors and industries have alliances, partnerships, forums, committees, subcommittees, study groups, working groups, liaisons, and various other interrelationships among themselves. (Once again, picture a plate of spaghetti with meatballs, or linguini with clam sauce, if you prefer.) This is even before you get to all the consulting firms that overlay all this organizational structure (a little bit like the grated cheese to top off the plate of pasta).

The consultants for complex systems will be quick to point out how many problems arise with complex systems and how many possible solutions there are to choose from. One of the biggest pitfalls in complex systems is the temptation to go off chasing solutions well before the problem is carefully defined and understood. This goes hand-in-hand with the temptation to chase solutions with lots of money; complex systems are typically very expensive. The payback period for money spent on complex systems is usually quite long, and those who make decisions to spend money at early stages of the project are rarely in the same job when the results of the expenditures, the return on investment, become clearly known.

With large, expensive, complex systems come high stakes and large risks. This is a key reason why few involved in the early stages stick around for the outcome. A reason why people prefer chasing solutions rather than carefully defining the problem space is that complex systems involve large numbers of requirements from many different stakeholder factions. The requirements often conflict, and the conflicts that show up in requirements are usually a manifestation of deep-seated conflicts between different parties, each an important stakeholder in the System.

While requirements are usually stated in writing, or at least verbally, complex systems must also take into account many constraints that are not often explicitly stated and sometimes not even consciously known. The constraints can be imposed by the primary organization responsible for the System, by any of the other, secondary organizations involved in defining and building the System, or by external factors such as legal, regulatory, financial, societal, or other forces.

7.3.2. The Role of Architecture in Coping with Complexity

Once a particular undertaking is recognized as a complex system, the central challenge is to properly structure the project, right from the very beginning. William Livingston's entire book, *The New Plague: Organizations in Complexity*, is devoted to explaining the consequences (which he refers to in descriptive language such as the "plague" and "wreckage") of trying to address complex problems using traditional approaches. The traditional approaches were defined for solving "simple" problems (i.e., those that can be fully understood by a single person), and in that role they are unparalleled in their effectiveness. But if an organization attempts to solve complex problems using these traditional techniques, the problem will "blow up in their face" (figuratively, if not literally).

The key step necessary to keep a complex problem, requiring a complex system for its solution, from winding up collapsing under its own weight, is to begin the project from the start with the formation of an architecture team. While forming the architecture team is necessary, it is not sufficient. To be successful, the team must have the right composition and must be given the right charter. Before we attempt to define the right composition of an architecture team, let us first look at what we mean by architecture.

7.3.2.1. What Is Architecture, and What Do Architects Do? Architecture is bringing together, for the purpose of defining a problem and its solution, various specialties, each representing a distinct body of knowledge and application, while resolving the conflicts that naturally arise between the differing viewpoints brought to bear by each specialty. Architecture is making all the pieces fit—integration.

One of the most difficult aspects of putting together an architecture team is understanding and describing what it is that architects do and how it is different from other responsibilities, such as systems engineering. A meaningful distinction is that architects specialize in reducing complexity, uncertainty, and ambiguity to workable concepts, while systems engineers specialize in making feasible concepts work.[12] Another distinction is that architecture focuses on *function*, while systems engineering focuses on *form*. Put another way, architecting is elucidating the *function* (or purpose) of a system so that systems engineering can direct its developers to construct and integrate the components that go into making up its *form (or structure)*.

Another way to distinguish the role of architects from other disciplines is to describe the characteristics that make for an effective architect. Some of the key characteristics that are *essential* for a good architect to possess are also some of the most difficult to find, almost running counter to the typical profile of an engineer:[13]

---

[12] *Systems Architecting*, p. 12.
[13] Ibid., p. 290.

- A high tolerance for ambiguity and uncertainty.
- An ability to work consistently at an abstract level but then to move to concrete details at the appropriate junctures in the life cycle.
- A willingness to backtrack, diverge, and seek multiple solutions.
- An almost childlike curiosity.
- An ability to relentlessly keep posing questions, following threads.
- A generalist's perspective.
- The ability to see issues from the perspective of various stakeholders, deriving the most salient aspects of the different positions, while seeking a balanced outcome.
- An ability to look at problems from a different, fresh perspective.
- An ability to think "outside the box."
- An ability to draw metaphors and analogies from far afield, bringing them to bear on the current problem space.
- Creative, obsessive juggling of requirements, constraints, technology.
- An ability to see things long before they exist (a.k.a., a sense of faith or vision), even in the face of strong opposition.
- A sense of mission and destiny larger than the current context.

A more concrete attribute of a systems architect is that the architect should have ample (e.g., at least ten years) professional experience. The architect must have broad knowledge of, and exposure to, all major areas of the company's business, and even the industry the company is in. In addition, the architect must also have experience in and exposure to several (at least three or four) of the primary technical disciplines in which the company has core competencies. The following excerpt speaks to this need for technical breadth, a quality that tends to be deemed less important in an era when specialization tends to be overemphasized:

> Knowledge breadth is just as important as technical depth, for it facilitates the early identification of potential problems and the integration of the activities performed by various disciplines. Outside of one's principal discipline, only a working knowledge is necessary: a basic understanding of how the choices one makes affect other people's work. For example, both design and manufacturing engineers should be able to predict the impact of a design choice on other elements in the system.[14]

In addition, the architect must have the confidence and desire to branch quickly and deeply into new technical areas. It is very important that an architect be able to learn enough about a new field in short order so that meaningful technical exchanges can be conducted with true experts in that field. The architect must be good at

---

[14] *Dynamic Manufacturing*, p. 300.

networking—first, knowing bright people in key areas of the company (and the industry at large), and second, being able to follow leads, or even make "cold calls" to rapidly locate vital information sources. The architect must recognize that vital sources of knowledge exist in the industry at large, quite likely crossing corporate and perhaps even competitive boundaries. As Bill Joy, of Berkely Unix and Sun Microsystems fame, has put it: "Most of the bright people don't work for you—no matter who you are. You need a strategy that allows for innovation occurring elsewhere."[15]

7.3.2.2. Architecture Team Composition.

As defined above, architecture involves bringing together various specialties, each representing a distinct body of knowledge and application. The most essential aspect of building an architecture team, apart from finding enough qualified architects, is covering enough of the core specialties that will be represented on the project. Sometimes this is difficult to ascertain a priori, and so the members of the team must at least collectively have a literacy in all of the necessary areas. Team members must be able to track down missing information very quickly.

The second aspect of architecture mentioned in our definition above was: *resolving the conflicts that naturally arise between the differing viewpoints brought to bear by each specialty.* Architecture inherently involves many conflicts. There are conflicting requirements. Requirements conflict with constraints. Desired features conflict with budget and schedule. But with large, complex systems, conflicts arise at a different level because of the various and diverse viewpoints brought to the table by the wide range of disciplines and expertise required. The following quote further elaborates on this issue:

> In working on complex problems there is a "linkage escalation" in having to include diverse groups, the lack of a suitable working context and a problem-solving process, organizational value-system conflicts, etc. In most cases, *the added complications of the resolution process are far more difficult to solve than the original problem.*[16]

There are several reasons why these added complications arise:

- Energy is dissipated and focus is lost when the different constituencies are allowed to fight over turf rather than focusing on common goals.
- In the absence of common goals, turf issues will rise to the surface; hidden agendas and motives, internal competition, and "baggage" will dominate.
- Existing solutions carry biases, perceptions, and egos that become entangled with the substantive technical issues.

[15] Quoted in George Gilder, "The Coming Software Shift," *Forbes ASAP*, August 28, 1995.

[16] *The New Plague*, pp. 1–13.

The architecture team must be more than just a group of representatives from their home departments. There must be a recognition of "superordinate goals"[17] or "overarching values"[18] that place the needs of the project, and its value to the company as a whole, ahead of parochial issues. This is easier said than done. If this does not take place, however, the project will suffer extensively.

> ... people involved tend to be relatively specialized and can be difficult to coordinate. The process is complicated and further politicized by the important tradeoffs that must be faced. The fragmented nature of the project team, composed of people whose primary allegiance is to a home department, specialty, or function, often leads to the perception that each trade-off decision results in a winner and a loser. Each group therefore tries to minimize its risk of losing, which tends to lead either to safe, noninnovative decisions, or to bureaucratic gridlock.[19]

Clearly, the architecture team needs to be composed of individuals who can technically and politically represent their constituent organizations, while at the same time identifying even more strongly with the central goals and objectives of the architecture team and those of the overall project life cycle. Toward this end, it is important that the team members be committed to the project for the duration, and not be pulled on and off the team, or required to participate on multiple parallel efforts simultaneously and fragmentarily.

Early on, the team members need to be adept at obtaining input from downstream stakeholders, at the proper time, and to the appropriate level of detail. The architecture team needs to begin getting the right level of buy-in from the downstream players early in the project life cycle. Later on, the architects need to be capable not only of feeding information to the downstream recipients, but of doing so in such a way as to gain a deeper level of buy-in. The architecture team members need to facilitate information-sharing between upstream and downstream stages at each stage in the life cycle to such an extent that potential conflicts are surfaced, and addressed, as early as possible.

Since we have already begun to cross over from addressing the architecture team composition (who they are) to addressing its charter (what they do), let us turn our full attention to charter.

7.3.2.3. Architecture Team Charter.
We have already touched on some of the roles the architecture team must play, including information gathering, conflict identification and resolution, stakeholder management, and information transfer. Now we need to define the role of architecture team more systematically.

---

[17] Richard Tanner Pascale and Anthony G. Athos, *The Art of Japanese Management: Applications for American Executives*, Warner Books, 1981, p. 125.

[18] Peter B. Vaill, "The Purposing of High-Performing Systems," *Organizational Dynamics*, Autumn 1982, p. 29.

[19] *Dynamic Manufacturing*, p. 312.

7.3.2.3.1. DEFINE THE PROBLEM, NOT THE SOLUTION. A natural inclination, which must be avoided at all costs, is for the architecture team to jump immediately to defining the solution to whatever the problem may be. While arriving at the solution is the very reason the architecture team, and moreover the entire project, exists, attempting to jump ahead to a solution for a complex problem is a sure formula for disaster. Instead, the first priority of the architecture team must be to adequately *define the problem*. As W. L. Livingston warns in *The New Plague*, there are far too many solutions looking for problems, and if the architecture team does not make a conscious and sustained effort to avoid it, the solution "hordes" will make every effort to force fit their solutions onto the architecture team's emerging (complex) problem, regardless of how good the fit really is.

Fred Brooks well captures the central role that the architecture team must play:

> The hardest part of building a software system is deciding precisely what to build. No other part of the conceptual work is as difficult as establishing the detailed technical requirements, including all the interfaces to people, to machines, and to other software systems. No other part of the work so cripples the resulting system if done wrong. No other part is more difficult to rectify later.[20]

The central difficulty with jumping too quickly to solutions before the problem is adequately defined and understood is that in the rush to arrive at solutions, the hard work of unraveling the essence of the problem can readily be avoided. By hastily attempting to solve the wrong problem (but one that appears to fit existing solutions), a tremendous amount of time and money can be committed early in the project to a path that leads ultimately to only disillusion and waste.

A characteristic of projects that fail is that mistakes made early on, once they move forward into the traditional project solution machinery, are typically only brought to light at the end of the project, at a disastrous day of reckoning. At that point, not only have lots of time and money been burned up, but attempts to recover only make the situation worse. As Livingston describes so thoroughly, the only way to avoid "The New Plague" is to get it right (i.e., focus on defining the problem), right from the start.

Brooks hits on related points:

> The most important function that the [architect] performs for the client is the iterative extraction and refinement of the product requirements. For the truth is, the client does not know what [he] wants. The client usually does not know what questions must be answered, and he has almost never thought of the problem in the detail necessary for specification.[21]

7.3.2.3.2. DISCIPLINED PROBLEM DEFINITION. Defining and characterizing the issues to be addressed, and the critical path for addressing them, is at the heart of what an architecture team must accomplish collaboratively and iteratively with the

---

[20] "No Silver Bullet," p. 17.
[21] Ibid.

client in architecting a complex system. Not only must the architecture team take great care to define the problem, but also they must unravel the problem in a systematic, disciplined way. At each step along the way, the team needs to be addressing what is viewed as the most critical issue in defining the problem, and they must also be aware of the next most important issue to be addressed.

Risk management is a key aspect of this systematic problem definition. The prioritization of critical issues must be tied to the risk (or uncertainty) associated with each issue. The level of detail to which the problem is analyzed, as well as the level to which risk-resolution strategies are defined, is a function of the associated risk. Issues involving greater risk are analyzed in more detail, while less risky issues are deferred until later in the project.

In defining the priority sequence for tackling issues to be addressed, the team must rely on their collective knowledge. Since the defining characteristic of a complex problem is that it exceeds the capacity of a single architect to understand, it is critical that the issue-sequencing process not be dominated by any one influential individual with some particularly strong bias that may seemingly, but not actually, bolster that individual's limited understanding.

7.3.2.3.3. DETAILED KNOWLEDGE OF SYSTEM BEHAVIOR (ASHBY KNOWLEDGE). The mission-critical task of the architecture team is to *create detailed knowledge about the behavior of the system under the anticipated range of operating conditions*—what Livingston calls "Ashby" knowledge.[22] The name is derived from *Lord Ashby's Law of Requisite Variety* which states that "*in order to be in total control [of a System] you must have the means to deal with whatever conditions the System may exhibit.*" In other words, "*the variety of threats existing in a System must be matched with an equivalent variety of suitable response capabilities.*" We will return to this topic subsequently.

7.3.2.3.4. ARCHITECTURAL INTEGRITY, SHARED MENTAL MODELS, COMMON LANGUAGE. As the architecture team begins to develop a solid understanding of the problem space, it is critical that they simultaneously begin to create a shared mental model and a common language of understanding. This is an indispensable tool in maintaining architectural integrity, first within the architecture team and later throughout the project life cycle. As Brooks points out, integrity of architectural concept is essential to good system design:

> I will contend that *conceptual integrity* is the most important consideration
> in system design. It is better to have a system omit certain anomalous fea-
> tures and improvements, but to reflect one set of design ideas, than to have
> one that contains many good but independent and uncoordinated ideas.[23]

Unfortunately, the architecture of a complex telecommunications management system-of-systems, unlike that of a building, a bridge, a circuit board, or other physical structures, can be quite difficult to visualize. And even if individuals form their own internal mental models of some portion of the overall architecture, con-

---

[22] *The New Plague*, pp. 4–61.
[23] *The Mythical Man-Month*, p. 42.

veying this clearly to others is a challenge. In addition, coming up with shared mental models that the entire team, and moreover, the project as a whole can jointly use to guide and direct their work cohesively and consistently, is the central challenge. To achieve this end, the architecture team must be very creative. Tools such as figurative language can be used as a powerful, creative mechanism for stimulating the collective imagination of the architecture team toward shared mental models. We will address this topic further in Section 7.3.2.3.5.

Let us briefly review why conceptual integrity is so important on a complex system project. The expertise needed to solve complex problems comes from many different organizations, representing specific expertise. When expertise from different disciplines is brought together, each brings to the table his or her own vocabulary and conceptual context. Unless the architecture team is able to knit together a conceptual framework on which all of this expertise can be brought to bear in a unified, cohesive fashion, the power of the expertise will quickly be dissipated. The architecture team must create a shared mental model—"pictures in their heads that are strikingly congruent"[24] to focus the energy and talent of the team on specific common goals and objectives.

To be useful, the output of the architecture team must be used to put the solution experts to work at the right time, and in the proper sequence, on solving the problems defined by the architecture team. So in addition to the shared mental model, the architecture team must also come up with a structured approach to translating the problem space into a solution space. The solution space must then be structured into pieces that can be farmed out to the solution teams. The shared mental model is critical to making this handoff effective, and moreover, to successfully implementing a complex system.

7.3.2.3.5. TOOLS FOR CREATING SHARED MENTAL MODELS, COMMON LANGUAGE. Tools such as figurative language can be used as a powerful, creative mechanism for stimulating the collective imagination of the architecture team toward shared mental models.[25] This is especially true during the intense learning cycles involved with creating Ashby knowledge, and most specifically, when attempting to translate tacit *operational* knowledge into explicit *conceptual* knowledge. Figurative language, such as metaphor and analogy, can be used to "find a way to express the inexpressible."[26]

Metaphor is "*a way for individuals grounded in different contexts and with different experiences to understand something intuitively through the use of imagination and symbols without the need for analysis or generalization.*"[27] This can be in the early stages of a project to begin blending a shared context among team members with varied perspectives and reference frames. What is intuitively obvious to someone from one background may mean nothing to someone from a different background.

---

[24] "The Purposing of High Performance Systems," p. 26.

[25] Ikujiro Nonaka, "The Knowledge-Creating Company," *Harvard Business Review*, November–December 1991, pp. 99–101.

[26] Ibid.

[27] Ibid.

The architecture team must begin to create shared knowledge by creating somewhat fuzzy concepts, expressed in carefully chosen metaphors, that begin to take on powerful meaning within the emerging, common context of the team. The power of metaphor comes in its ability to establish a connection between two different, separate areas of experience, and thereby to set up a mental conflict due to the inherent discrepancy or conflict between the two ideas. The architecture team can thereby harness the energy bound up in the inherent conflict captured by the metaphor as the team collectively wrestles with the deeper, richer insights embodied in the metaphor.[28]

For example, Nonaka describes the process by which the design team at Honda came up with a radical new car design for the urban market. The metaphor used was "Theory of Automobile Evolution." The inherent conflict between combining the *automobile*, an inanimate object, and *evolution*, a theory dealing with living organisms, into a single concept, sparked the team to create a new way of looking at automobile design. They came up with the "Tall Boy" concept, which is both "tall" and "short," and fits better than traditional car designs with the needs of a human in an urban setting to have ample space inside the car, while having the car present a small footprint. This launched a new era of car design in Japan, based on a man-maximum, machine-minimum concept.

Analogy provides a further refinement of the conceptual tensions created by the metaphor. Analogy takes the abstract differences embodied in the metaphor and translates them into more concrete images. By using carefully chosen analogies, the conflicts involved in the metaphor begin to be resolved through a more systematic analysis of how the two ideas in the metaphor are, and are not, alike. In the case of the Honda Tall Boy, a key analogy was the image of the car as a sphere. This concept provided some breakthrough thinking by allowing the car to be both tall (in height) and short (in length). The sphere enabled a design that was lighter, cheaper, more comfortable, and more solid than traditional car designs. Of course, the ultimate design modified the spherical shape in important ways to address practical engineering and operations considerations, but it was a powerful analogy for helping the team to leap ahead in refining their thinking.

7.4. THE ROLE OF LEARNING AND MEMORY IN COUNTERING COMPLEXITY

Organizations that have become successful at building simple systems (systems that are solutions to simple problems, whose architecture can be completely defined by a single person) have developed expertise encoded in certain organizational practices. As long as the organization is attacking simple problems, the established practices tend to continue to work very well, with only minor adjustment required from time to time. An organization engaged in gradually improving its established practices for

---

[28] Ibid.

addressing simple problems can be described as being engaged in "single-loop learning."[29] Single-loop learning is the most basic and most common type of learning for individuals as well as for organizations.

When an organization with established practices for addressing simple problems is confronted with a complex problem that it must address, the strong natural tendency is to attack the problem using the same techniques and practices that have been refined for addressing simple problems. Attempts will be made to use single-loop learning to correct the practices to adapt to the complex problem. However, the practices are ill-suited and are downright inappropriate for solving complex problems, at least during the critical, up-front stage of thoroughly defining the problem.

If the project is embarked upon using the established practices, then the project is destined for failure. The only way the project can tackle the complex problem, and come up with a true solution, is for the organization first to undergo "double-loop learning" (which we will explain subsequently) at the outset of the project. However, as Livingston points out, shifting from the ingrained approach (single-loop learning) to the required approach (double-loop learning) is quite difficult because it requires an "*institutional correcting mechansim*"[30] that is not typically or readily available.

After some background is provided on theories or models of learning, the distinction between single-loop and double-loop learning will be explained in detail, as will their relationship to architecting complex systems. As we will see, complex solutions that are truly responsive to the mandates of complex problems can only be architected by organizations adept in knowing when and how to switch from single-loop to double-loop learning.

7.4.1. Individual Learning and Memory

Before we delve into a discussion of organizational learning, we first need to lay some background on individual learning and memory. As we will see, organizational learning and memory is modeled as an extension and an abstraction of individual learning and memory.

7.4.1.1. Models of Individual Learning. For our purposes, individual learning is defined as: *increasing an individual's capacity to take effective action in the future, through enhancing both skill and knowledge based on past experiences.*[31] Two important ingredients in this formula are (1) skill, or *know-how*, which implies the physical ability to produce some action, and (2) knowledge, or *know-why*, which implies the ability to articulate a conceptual understanding of an experience. The significance of these two facets of learning cannot be under-

---

[29] Chris Argyris, "Double loop learning in organizations," *Harvard Business Review*, September–October 1977.

[30] *The New Plague*, pp. 5–65.

[31] Daniel Kim, "The Link Between Individual and Organizational Learning," *Sloan Management Review*, Fall 1993, p. 38.

stated, and as such the relationship between know-how and know-why will be further explored in section 7.5.

Various models of learning have been put forth which address the relationship between these two basic aspects of learning (skill and knowledge). The school of thought that seems to work best is called experiential learning theory. Experiential learning is described as a cycle with four stages, two of which focus on operational (know-how) learning, and two of which focus on conceptual (know-why) learning. Readers familiar with TQM (total quality management) will recognize this cycle as plan-do-check-act (PDCA). Other researchers have used different nomenclature to describe essentially the same model.[32] The model we will adopt is called OADI—observe-assess-design-implement—and is shown in Figure 7.3.

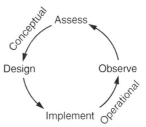

Figure 7.3 Model of Individual Learning.

Observation involves experiencing concrete events and actively perceiving what is happening; *assessement* involves reflecting on the experience; *design* entails constructing conceptual models based on the assessment; and *implementation* entails testing the design by performing some set of concrete actions in the physical world, such as building a prototype. The last step leads to a new concrete experience, which begins the cycle again. As an aside, this basic learning cycle, drawn from experiential learning theory, is the foundation for the spiral model of software development advocated by Barry Boehm,[33] which is touted as a "risk-management" approach to managing the software development life cycle.

7.4.1.2. Models of Individual Memory. Just as we recognize the indispensable role that memory plays in our own individual learning experiences, so it is important to recognize the role of memory in a conceptual model of learning. As we will see, memory plays a key role in linking our model of individual learning to that for organizational learning. As with the model for learning, the distinction between know-how and know-why emerges as an important aspect of the model for memory.

For the purposes of our discussion, we will assume a basic relationship between learning and memory: *learning pertains to acquiring information to be placed in mem-*

[32] Ibid., pp. 38–40.
[33] Barry Boehm, "A Spiral Model of Software Development and Enhancement," *Computer*, May 1988.

ory; memory pertains to retaining information acquired via learning. However, we must recognize that there is a feedback mechanism in that what is already stored in memory affects what we learn.

It is helpful at this point to return to the concept of mental models. Earlier, we discussed the idea of a shared mental model. Our individual mental models operate as an active substrate, or filter, within our memory that colors and shades not only what we observe and experience, but also how we interpret and act upon our observations and experiences. These mental models can be compared, by way of rough analogy,[34] to the source code of an operating system, managing and arbitrating how we acquire, retain, use, and delete new information.

Figure 7.4 shows the relationship between individual learning and individual memory. Following on the work of several researchers in the field,[35] we take individual memory to consist of two parts, corresponding to the operational and conceptual types of learning. Daniel Kim calls these two forms of memory *routines* (operational—know-how) and *frameworks* (conceptual—know-why). Others have called this *procedural* and *declarative* memory, again drawing an analogy to computer programs:

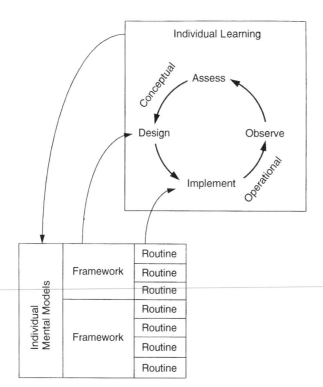

Figure 7.4 Model Relating Individual Learning and Memory.

[34] "The Link between Individual and Organizational Learning," p. 39.
[35] Ibid., p. 40.

... which [computer programs] may exist as compiled machine language (procedural), or as high level language source code (declarative). The former is rapidly executable, but difficult to repair and closely tied to a specific hardware environment. The latter can be repaired or generalized to other environments more easily, but can be executed, in a typical case, only by a very slow interpretation process.[36]

Operational (or procedural) learning involves performing particular skills and tasks that are governed by existing, encoded routines. However, these routines may also be updated, modified as a result of operational learning. Routines "are the parts [of an individual's skill] that he or she does not have to think about—once a routine is switched on in the worker's mind, it goes on [to] the end without further consultation of higher faculties."[37] When faced with an "interrupt" (i.e., an unusual or unexpected situation) in performing a routine, an individual typically does one (if not both) of two things: (1) invokes a different, existing routine to handle the situation, or (2) modifies the currently operative routine to accommodate the situation.

Routines can be thought of as being nested "inside" frameworks, with a specific set of routines supporting a given framework. A single framework may have dozens, and perhaps even hundreds, of routines associated with it. A framework forms the conceptual, scientific, theoretical underpinning for a family of routines. For example, a repair technician may have a framework for understanding SONET as a technology, including how individual ADMs (add/drop multiplexers) can be configured in different architectures (rings, chains, hubs, etc.). Within that framework, the technician then learns a set of routines that indicate what actions to take on individual ADMs when specific faults occur in a particular SONET architecture. The technician would have a different framework for NGDLC (next-generation digital loop carrier), as well as different routines. However, the frameworks, as well as the routines, would have some overlap since the SONET and NGDLC technologies share some common aspects.

7.4.1.3. Individual Single-Loop Versus Double-Loop Learning.
We now need to look at how the two different types of learning (single- and double-loop) take place, using our model that relates individual learning and memory. Figures 7.5 and 7.6 show how single-loop learning, and double-loop learning are similar and different, and how each is related to the individual learning/memory model.

First, let us provide the following working definitions of single-loop and double-loop learning, and then we will look at some examples.

---

[36] Michael D. Cohen, "Individual Learning and Organizational Routine: Emerging Connections," *Organization Science*, February 1991, p. 137.

[37] Ibid., p. 135.

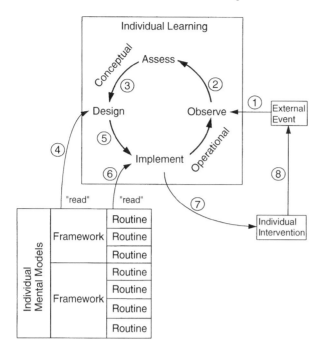

Figure 7.5 Individual Single-Loop Learning.

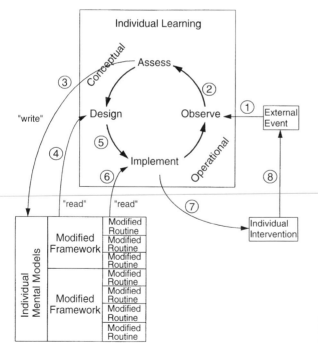

Figure 7.6 Individual Double-Loop Learning.

Single-loop learning (SLL): detecting and attempting to improve the operational efficiency of normal practices and procedures (routines) so that established mental models (frameworks) can be maintained and incrementally improved.

SLL responds to an external event by invoking the OADI cycle, and then executing some individual intervention based on existing frameworks and routines, or at most, slightly modified routines within an existing framework. Put another way, SLL is "read-only" on frameworks, and mostly "read-only" on routines, with an occasional, minor "write" performed on a routine.

Double-loop learning (DLL): detecting and attempting to correct basic flaws in the established policies and objectives (frameworks) underlying practices and procedures (routines), so that the frameworks can be revised and new or improved routines can be established.

DLL responds to an external event by performing observation and assessment, and then updating frameworks and routines. These new frameworks and routines are then used as the basis for implementing some new individual intervention based on the new frameworks and routines. DLL performs a "write" on a new or modified framework, and "writes" new or modified routines within these frameworks.

Thus, single-loop learning makes only minor changes to routines, while striving to preserve frameworks, whereas double-loop learning seeks to overhaul existing, or establish new frameworks, causing the need to create new or significantly modified routines as well.

7.4.1.4. An Example of Individual Single-Loop Learning.
In the case of single-loop learning, an individual is made aware of an environmental situation. For example, a repair technician is provided with a dispatch notification, indicating that a specific fault has occurred at a particular location in a SONET subnetwork. The technician has received training which established a conceptual framework that includes a specific set of routines for repairing various types of faults in the SONET equipment. Based on this training, the technician executes a series of routines that ultimately result in the fault being corrected and the dispatch being closed out. The key point is that the existing routines within the existing frameworks are adequate to address the situation. At most, the specific routines may be adapted slightly or refined to correct minor discrepancies between what the technician was taught in the training courses and the way things acutally work in the field.

7.4.1.5. An Example of Individual Double-Loop Learning.
Now suppose that this same technician, over a period of time, observes that when a particular type of condition occurs, even after all of the correct routines have been executed, the problem still exists. The technician would then begin to question specific routines,

and to some extent the framework for addressing this condition. This might trigger the technician to perform some additional analysis, and begin to develop a new or modified framework (mental model) for how the SONET equipment behaves, particularly in this type of situation. In addition, the technician would devise some specific routines for resolving the condition.

The key in this example of DLL is that the environmental situation (or series of situations) caused the technician to question the existing framework and routines, and then begin investigating significant revisions or expansions to the framework and routines. This is the essence of double-loop learning. Whereas the single-loop learning was *corrective* (adjusting, fine-tuning existing routines), the double-loop learning was *generative* in that a modified framework and new routines were created. It is likely that in a DLL situation, a technician would actually seek outside help either within his own company (e.g., call the Service Assurance field support hotline) or from the equipment supplier's support hotline. This may result in a learning exchange, along the lines of those described in section 7.5.3.

7.4.2. Organizational Learning and Memory

Before we delve into an explanation of the organizational model of learning, as an extension of the individual model of learning, let us first set the stage. The literature is replete with studies of organizational learning, and there are many aspects to this very broad subject area. What we present here is drawn from a sampling of literature in this area, but with a specific focus in mind. We wish to focus specifically on the type of learning that is needed when an industry is in a tremendous state of flux and turmoil. This is the situation that many manufacturing organizations, such as companies in the automotive, steel, and electronics industries, faced over the last decade. As such, we can draw on their experiences.

Our starting premise is that an organization has a set of shared mental models that are ingrained in its structure. These models affect everything an organization does, both explicitly and implicitly. The models are often very difficult to articulate, although most people in the organization know what the models are, at least tacitly. In some cases, the models are codified explicitly into written procedures. These models govern both day-to-day decision making and less frequent decisions, such as what TMSs to purchase and how to interconnect them.

Keep in mind Conway's Law: "*Organizations that design systems are constrained to produce designs which are copies of the communications structures of these organizations.*" A logical conclusion from Conway's Law is that the architectural framework possessed by the TMS architecture team for an organization is directly tied to the shared mental models of the organization as a whole. To put it another way, the TMS architecture team will tend to make decisions based on their common understanding of how their organization interacts. Alternative ways of stating Conway's Law include:

> Organizations and their systems are symbiotic—each needs and must fit the other.[38]

> It is easier to match a system to the human one it supports than the reverse.[39]

To which we add:

> To affect a change in TMS architecture will require a prior change (or at least a simultaneously coordinated change) in the organization structure and dynamics.

Based on what we said earlier, the TMS architectural framework that resides in the organization, and is the basic blueprint for structuring the TMSs that the organization buys or builds, is a cross product of at least several key business forces:

- Market forces
- Business forces
- Network technologies
- Organizational structure and dynamics
- TMS (Telecom Management Systems) technology

If these forces are *relatively* stable and predictable, as they have largely been over the last several decades, then the architectural framework remains solid, pervasive, and is firmly operative in guiding all TMS architectural decisions. The architectural framework pervades all solution work on all subcomponents of the meta-system architecture. The problem comes when too many factors are changing at once:

> In other words, the extent to which members can come to share pictures in their heads about the organization's basic purposes depends on some degree of stability in (1) environmental demands, (2) members' own expectations and needs, (3) the technologies they are operating, and (4) the structures through which they are bound together. *Beyond some unknown threshold, too much change in this system of factors breaks down the shared sense of what the organization is, why it exists, and what its basic purposes are.*[40]

In other words, beyond some threshold, the conceptual integrity of the architectural framework used by the TMS architects will break down. In our discussion we are interested in determining how learning occurs under these conditions. How does an organization decide when to change its TMS architectural framework (perhaps as things begin to break down), and how does this transformation take place once such a decision is made?

---

[38] *Systems Architecting*, p. 270.
[39] Ibid., p. 136.
[40] "The Purposing of High Performance Systems," pp. 28–29.

7.4.3. Organizational SLL and DLL

Just as with individuals, organizations can undergo single-loop and double-loop learning. In these terms, what we are really interested in is organizational double-loop learning. This is not to undermine the significance of single-loop learning in organizations. Some have argued that "improving speed of [organizational] routines and changing their detailed contents, along with accurate switching among existing routines, are major sources of competitive advantage or other forms of organizational success."[41] This type of learning amounts to incrementally improving how an organization operates and is really akin to removing accidental complexity. This is an important aspect of organizational learning, but it is not the central focus of our current discussion.

One way to think of the significance of organizational double-loop learning is to look at the results of ignoring the need to invoke DLL to address complex problems. To do this we briefly examine the role of bureaucracies. A bureaucracy represents an efficient means of organizing large numbers of people into a highly efficient and effective "machine." A bureaucracy can be viewed in a positive light as an organization that excels in SLL. Such a company has developed well-established business practices involving frameworks and routines built up incrementally over many months and years. These practices are unparalleled in managing business endeavors *where past experience is an accurate predictor of future activity*:

> Bureaucracies are best able to handle . . . structured, stable, and rational situations. They are least able to handle novel, . . . unstructured, dynamic, and irrational situations for which the rules have not yet been written. For what they do best, they are unexcelled.[42]

For organizations with bureaucratic tendencies, one can think of organizational double-loop learning as the mechanism by which nonbureaucratic business methods are created in the face of business endeavors where past experience is *not* an accurate predictor of future activity. This involves what Peter Drucker calls "the most difficult learning imaginable: unlearning."[43] Unlearning is required because the existing set of frameworks and mental models by which all organizational behavior is governed must be brought under question and close scrutiny. Fundamental assumptions must be brought to the surface and dissected with seemingly irreverent abandon. It is neither an easy nor a popular undertaking.

Once again, an organization that has perfected SLL on a large scale has essentially become a bureaucracy. This is acceptable unless the organization undergoes a climate change in which multiple, significant factors are all changing simultaneously—industrial strength complexity. Under these circumstances, the bureaucracy

---

[41] "Individual Learning and Organizational Routine: Emerging Connections," p. 136.

[42] *System Architecting*, p. 271.

[43] Peter Drucker, "The New Society of Organizations," *Harvard Business Review*, September–October 1992, p. 102.

does one of two things: It either engages in DLL, or it ignores the need to do so. Organizations in this latter state tend to focus more and more energy on internal issues and ultimately on fighting internal rivalries. The result is that less energy gets spent fighting competition and focusing on external issues and the needs of customers.

7.5. THE SIGNIFICANCE OF COUPLING OPERATIONAL AND CONCEPTUAL LEARNING

The distinction between *know-how* and *know-why*, or *theory* and *practice*, is something we are all familiar with. However, we may not be aware of just how significant the relationship between these two types of knowledge is with regard to the sustained success of organizational learning, especially in the face of adversity and turmoil. Before we explain the significance of properly relating know-how to know-why, we need to get a handle on the diversity of language that is used to describe these two aspects of learning. Table 7.1 illustrates the many synonyms used variously to express these two concepts. The last two entries are some characteristics of these two types of learning.

TABLE 7.1   SYNONYMS TO EXPRESS KNOW-HOW AND KNOW-WHY

Know-how	Know-why
Operational	*Conceptual*
practice	theory
tacit	explicit
implement	plan
craftsperson	analyst
doer	thinker
implementor	planner
procedural	declarative
routines	frameworks
line	staff
artisan	scientific
experiential	observational
concrete	abstract
shop floor	ivory tower
hands-on	blue-sky
manufacturing	engineering
not readily articulated	readily articulated
not readily transferable	readily transferable

In our models of individual and organizational learning, two distinct aspects of learning were evident: operational learning and conceptual learning.

Operational learning is conducted at the beginning of the learning cycle via *observational* learning, and at the end of the learning cycle via *implementation*. Operational learning is primarily captured in *routines*, which are detailed, step-by-step methods for performing specific tasks. Conceptual learning can be viewed as the middle of the learning cycle, being fed by (hands-on) observation at the front of the learning cycle, and feeding (hands-on) implementation at the end of the cycle.

Conceptual learning consists of assessment and design, and is captured in *frameworks*, which are models and higher level end-to-end process flows that tie together and integrate a complete set of routines required to perform a business process.

7.5.1. An Historical Perspective

The importance of balancing these two aspects of learning cannot be emphasized enough. Extensive sections in the book *Dynamic Manufacturing: Creating the Learning Organization* are devoted to explaining the significance that this balance, or the lack thereof, has played in American manufacturing over the last century. Much of what is expounded in that text is applicable to the TMS life cycle as well.

> The key to American preeminence had been the meshing of the skills and traditions of the craftsman with the thought process of the scientist, *the combination of experience and analysis, the mixture of systematic procedure and individual artistry*. Harnessing art and science together was powerful because the two traditions helped to offset the weaknesses inherent in each.

> That balance was a precarious one, however. . . . *About the only way an organization can maintain this uneasy alliance is through a central core of people who have been trained in both*. Once this binding core is lost, and as one side begins to gain power over the other, no countervailing force is likely to assert itself to reestablish that balance.

For TMSs, an imbalance slanted toward the operational side could be manifested in what we earlier termed "the tyranny of the SMEs." This refers to the tendency for TMS users to control the budget for system enhancements, choosing incremental enhancements offering near-term, task-oriented improvements, rather than funding more fundamental modifications that require a change in business process, a shift from "the way things have always been done."

A bias in the other direction, in favor of conceptual expertise, might manifest itself in the TMS arena in a set of powerful but complex system features that, while representing "breakthrough thinking," are not used by the system users. The features are not being used because they are complicated to learn at a time when the user community's training budget is crunched, or the features could be disregarded because they don't take into account realities in the way the users actually do business (e.g., due to constraints imposed by the labor union).

The authors of *Dynamic Manufacturing*[44] explain how this balance of "Old World" craftsmanship was wedded to scientific knowledge around the turn-of-the-

---

[44] *Dynamic Manufacturing*, Chapter 2.

century and how it led to American preeminence in manufacturing during the first half of this century. The artisan heritage had been brought to the United States through immigration from Europe, where the "learn by doing" method involving lengthy apprenticeships was used for centuries. Around the beginning of the twentieth century, engineering was just beginning to be introduced as a formal area of study into institutions of higher learning in this country. By joining the structured, analytical methods of science and engineering with the rich heritage of the craftsman tradition, a powerful combination was created.

A significant aspect of this union was the ability to gather observations (know-how) directly from the shop floor, analyze it using scientific methods (know-why), design focused experiments, perform the experiments, and then record observations to drive another cycle of learning. These types of experiments provided new, systematic insights into how business processes work in detail. With this experimentation came a methodical approach to investigating and then actually tracking changes to established procedures. As the "engineering" types came up with new ideas about how to improve the *routines* of the "manufacturing" types, these routines could be tested and refined through experimentation. At the same time, the larger conceptual *frameworks* being postulated by the theoreticians to govern larger business processes could also be tested and refined. The following quote illustrates the combination of "engineering" and shopfloor expertise working in unison:

> Art Raymond (1951) describes how the DC-1 was conceived and built; engineering was in the loft and the shop was on the ground floor. All the engineers were in continuous contact with those building the planes. Close proximity of all parties meant free and informal communication.[45]

The powerful combination of artisan and scientific expertise worked extremely well for much of this century. The results of systematically combining know-how with know-why prove quite effective as long as the chain of communication and the cycle of learning are kept well intact. However, an important phenomenon that can throw this combination out of balance occurs when the people on the front line (shop floor) start to have more ready access to cutting-edge information and when this information is changing at an accelerating pace. In these situations, such as we currently face in the telecommunications industry, the people in the staff planning roles can readily become too out of touch with the day-to-day, dynamic, vital essence of the business. The following sums up this point quite well:

> As organizations . . . become larger and more complex, the [people] at the top (whether managers or analysts) depend less and less on first-hand experience, more and more on heavily "processed" data. But what does the information processing system filter out? It filters out all sensory impressions not readily expressed in words or numbers. It filters out

---

[45] *Systems Architecting*, p. 280.

those intuitive judgements that are just below the level of consciousness. So the picture of reality that sifts to the top of our great organizations and our society is sometimes a dangerous mismatch with the real world.[46]

To a great extent, the breakdown of bureaucratic management is a natural result of the principles of scientific management, carried to their extreme, logical conclusion. Frederick Taylor, the most widely known promoter of scientific management principles, advocated the creation of specialized staff groups (possessing conceptual "know-why") to establish the processes, procedures and routine, of workers.[47] Line workers (possessing operational "know-how") specialized in individual work tasks that were governed by instructions, procedures, and assignments received from the staff groups. If not managed carefully, this can create a "me think, you do" kind of separation between "staff" administrators and "line" workers.

The danger is that if the distinction between conceptual learning by staff and operational learning by line personnel are not systematically and continually linked, then the work of the staff planners will drift away from the actual activity on the shop floor. Problems will arise (call them missed opportunities for greater, more meaningful conceptual learning) that the staff planners did not anticipate and were not prepared to address.

Unless the staff is consciously involved in the day-to-day, hands-on experiences of the line workers, lack of attention to continual improvement in worker skills, procedures, methods and routines will result. Ongoing, active feedback from the "factory floor" into the strategic planning process is vital to the successful evolution, and even survival, of the business.

What is the connection to the TMS life cycle? To deal effectively with the increased rate at which market forces, business forces, network technologies, and organizational structure/dynamics are changing the requirements for TMSs, it is vital that the TMS planners maintain close working relationships with operational expertise on the front lines.

Without actively working to maintain these ties, the effective transfers of operational to conceptual knowledge (and the corresponding reverse transfers) needed to keep the TMSs in step with the changing needs of the business will not occur. Without deliberately working to maintain these ties, the staff will naturally drift by tending to focus on problems that are of most interest to them, however potentially esoteric. Attempts to revamp business processes and to restructure the TMSs in kind are doomed to failure if the proposed solutions are not founded on operational knowledge obtained directly from the hands-on experts.

[46] John W. Gardner, quoted by Charles G. Arcand, "Bureaucratic Innovation, The Failure of Rationality," *Chemtech*, December 1975, p. 712.

[47] *Dynamic Manufacturing*, Chapter 2.

7.5.2. The Importance of Transferring Operational Knowledge

In many modern companies, especially in the telecommunications industry, an unwritten, unspoken priority seems to be given to conceptual knowledge and learning as compared to operational knowledge and learning. It is often built into the pay scale and the reward system. Part of moving up in a typical network provider is to rise from the ranks of the "line" organization to the loftier "staff" positions. A natural separation (or perhaps segregation) tends to arise between the two types of learning.

Apart from any social implications, the long-term adverse effect of this separation of know-how from know-why is that the deepest and richest aspects of learning become unavailable to companies that knowingly (or unwittingly) practice such behavior. Specifically, for companies involved in building or procuring TMSs, there are several areas where the systems architects and systems engineers (primarily possessing know-why) can "partner" more effectively with other members of the "team" at large who possess specific operational expertise (know-how).

Perhaps the most obvious experts with which systems engineers can engage in more complete learning are the software developers. In addition, however, the software testers and deployment support personnel are important sources of operational knowledge. For specific expertise in managing network technologies, the personnel in the *Technology Support* organizations (providing centralized support for field technicians working with specific network technologies) of the network providers are an excellent source of know-how. And, of course, the users of the TMSs in the network providers are also excellent sources of hands-on knowledge.

Members of *the TMS architecture team must cultivate ongoing relationships with experts possessing various types of specific operational knowledge.* Only by maintaining such prolonged relationships can the architects have ready access to tacit insights, intuitions, and "hunches" acquired only through operational learning. This source of knowledge is invaluable especially when the team is faced with a new, complex architecture for which the *Ashby knowledge* (mentioned in section 7.3.2.3) needs to be created.

As we will recall, Ashby knowledge is detailed information characterizing the behavior of a system under both intended and unintended circumstances. An architecture team defining Ashby knowledge for a new, complex system may, in some cases, have the luxury of involving experts with the right types of know-how during the development of this Ashby knowledge. However, since this may not be the case, it is critical that the team members be continuously acquiring operational knowledge so that this knowlege is at the team's disposal when needed.

Creating Ashby knowledge is an *artform* requiring that conceptual learning and operational learning be carefully balanced. Furthermore, once Ashby knowledge is developed for a complex system, this knowledge must ultimately be transferred into a new framework (or sets of frameworks), and new routines within each framework. At this stage, it is essential that the architecture team understand how the downstream organizations best adopt new technologies and practices. This includes how the existing frameworks and routines are implemented, recognizing *official* routines and frameworks, as opposed to the *unofficial* frameworks and routines that are actually in use. The architecture team must understand how to conduct the effectual transfer of both conceptual knowledge (embodied in *frameworks*) and operational knowledge (embodied in *routines*).

We have provided some motivation for tapping into operational knowledge. However, even if this need is recognized, and even if specific sources of this knowledge are identified, this does not necessarily make the ensuing task easy. The inherent difficulties in tapping operational knowledge are undoubtedly a reason why this type of knowledge transfer is not routinely practiced. Know-how is difficult to articulate, communicate, and formalize. It is embodied in the act of performing a task, and performing a specialized task consistently well typically requires years of hands-on experience to develop. Operational knowledge is captured in routines that become second nature to the expert, while perhaps at least initially confounding the novice. This type of knowledge is fundamentally difficult to capture and transfer.

Some of the techniques that can be used to enable the more effective transfer of operational knowledge revolve around a basic concept—creating a "common cognitive ground."[48] One of these techniques is to deliberately create overlapping responsibilities, activities and information among the assignments given to employees. Another, related technique is to strategically rotate employees through assignments that expose them to different areas, functions, disciplines, and technologies within the company. This type of cross-breeding of knowledge is so important because the way people interpret new information is so dependent on their current situation and perspective. Unless shared contexts are created across working groups within an organization, knowledge tends to change or lose its meaning when communicated to people with a different context.

7.5.3. Transferring Operational and Conceptual Knowledge

If we want to take fuller advantage of operational knowledge and integrate it with conceptual knowledge to create a more powerful learning cycle, we must look at how both operational and conceptual knowledge can be accessed and transferred. Quite simply, *four types of knowledge transfer are possible*:[49]

- Operational to operational
- Conceptual to conceptual
- Operational to conceptual
- Conceptual to operational

[48] "The Knowledge-Creating Company," p. 102.

[49] Ibid., pp. 98–99. Note that Nonaka uses the term *explicit* where we have used the term *conceptual*, and the term *tacit* where we have used the term *conceptual*.

7.5.3.1. Operational to Operational (Socialization). This type of knowledge transfer involves "learning by doing," hands-on, on-the-job-training. Perhaps the best example is the traditional apprenticeship method. Operational knowledge in this case is transferred by way of observation, imitation, and repetitive practice. A set of skills (i.e., one or more routines) is explained and illustrated by an expert, who imparts to a trainee the techniques and mental understanding necessary to perform these skills. Aspects of this learning can occur in the classroom or in a lab, but for the most part it occurs "in the field" (i.e., in a live, operational environment, on-the-job).

A limiting factor involved in operational knowledge transfer is that the emphasis is on transferring know-how, and not know-why. Through this type of learning alone, neither the expert nor the trainee gains much insight into the conceptual framework behind the routine. In addition, this type of knowledge is typically transferred from one master to one apprentice, or a relatively small number of apprentices. Finally, because this type of knowledge transfer does not bring the operational (tacit) knowledge to a conceptual (explicit, or expositional) level, the operational knowledge cannot easily be leveraged for use by the organization at large.

7.5.3.2. Operational to Conceptual (Articulation). This is the process of converting knowledge from know-how to know-why. This transformation is difficult to do well, because operational knowledge is knowledge embedded in action and is not readily expressed explicitly in written or spoken language. There is always the risk of oversimplifying the know-how, or not representing its entire essence. This is a major reason why the promise of developing expert systems has proved so elusive. Those familiar with the time and effort involved in interviewing experts for this purpose understand how difficult it is to extract, let alone codify, the expert knowledge.

The operational to conceptual knowledge transformation is best made by those who have actually learned to perform the routines themselves (via socialization and apprenticeship) and based on their first-hand experience can express the almost inexpressible. For example, an analyst can become an apprentice to an expert artisan, and then the analyst can capture the operational learning in an explicit form that can be more readily transferred as conceptual knowledge to other analysts or systems engineers. Only by extracting operational knowledge and making it available to a wider audience in the form of conceptual knowledge can the richer value embodied in the encoded routines be brought to bear in a broader domain.

7.5.3.3. Conceptual to Conceptual (Combination). This type of knowledge transfer is the type that most of us (telecommunications systems engineers) tend to be engaged in most often. This occurs when you read a technical memorandum or journal article and assimilate the information into your own mental model. Transferring conceptual knowledge is very much a "paper exercise" or "thought process," which is not to underestimate its power but to contrast it with the other two types of knowledge transfer.

The most significant power from transferring conceptual knowledge occurs when the conceptual knowledge being transferred is first derived from operational knowledge (existing routines). Then after the conceptual knowledge is transferred, it can be combined with other conceptualized routines and frameworks, and formed into new and expanded frameworks. Within these frameworks, new routines can then be conceptualized and ultimately translated back into operational knowledge.

7.5.3.4. Conceptual to Operational (Internalization).

Expressions such as "you can't teach an old dog new tricks," "set in their ways," and "that sounds easy in theory, but it will never work in practice" come to mind when considering the conversion of conceptual into operational knowledge. Operationally oriented individuals have a desire to want to "kick the tires" before they buy into a new concept. To a large extent, this type of reaction is warranted since the operational folks are the ones who will have to live with whatever new innovation is being foisted on them on a day-to-day basis for many years.

Distributed network management is a case in point. The concepts have been articulated and improved over the last five to seven years in various forms. The conceptual knowledge has been widely transferred. Yet progress in converting these ideas into practical, implementable products and methods has come slowly. The "thinkers" readily bought into the concepts, but the "doers," including software system development and operations support staff experts, have been less convinced.

OSI/CMISE is another case in point. This technology has been widely articulated for well over ten years. Prototypes and even a few production systems have been built. Yet, the fundamentally difficult task of taking this technology all the way down to the lowest level required for full implementation in a major network provider is only now being worked out in detail. The detailed methods, procedures, and routines needed by the end-user implementers are still under development. The "new kid on the block" always has the burden of fully displacing the reigning champion. The end-users do not want to adopt a replacement technology when the day-to-day utility of the new technology does not yet fully match (let alone exceed) the technology being replaced.

Of the four types of knowledge exchange described here, converting conceptual knowledge to viable, implementable operational knowledge is by far the most challenging and the most time consuming. Often, a "compelling event" or even a crisis is required to overcome the inertia involved in bringing a new conceptual framework to fruition. In part, this is based on financial realities. In part, it is based on the staid wisdom "if it ain't broke, don't fix it." But mostly it is because converting conceptual to operational knowledge is fundamentally taxing work. As Niccolo Machiavelli put it so eloquently in *The Prince*: "There is nothing more difficult to take in hand, more perilous to conduct, or more uncertain in its success, than to take the lead in *the introduction of a new order of things*."

7.5.3.5. A Knowledge Transfer Learning Cycle.

Each of these types of knowledge transfer is important by itself, and is a necessary part of crafting TMSs. However, to deal with the type of industrial strength complexity that is the central focus of our discussion, there is a need to sequence these different types of knowledge transfer in a specific order. A very powerful type of learning cycle occurs when these four types of knowledge transfer are sequenced in the following order:

1. *Socialization*: Operational knowledge, in the form of established but informal routines, is passed from a hands-on craftsperson to a systems analyst, systems engineer, or architect who acts as an apprentice. This operational knowledge resides in the head (and hands) of a field technician, a "tier 2" operations analyst, a highly experienced system user or database analyst, and so on.

2. *Articulation*: The systems analyst translates the operational knowledge into conceptual knowledge, formalizing the routine so that it can be communicated to other systems analysts (i.e., systems engineers, architects).

3. *Combination*: The analysts combine this formalized routine with other routines to form a larger, expanded framework. This could be the seed for a new set of business flows through an existing or modified systems architecture, or the genesis for an entirely new systems architecture. The framework and routines are communicated by the analysts to other stakeholders in the organization, and some concrete action steps are planned (e.g., building a prototype).

4. *Internalization*: The framework and routines are tested in some sort of experiment, trial, or prototype by executors of routines, and in the process new routines are created and the existing framework is enlarged.

The key to this process is that someone of a "scientific" (conceptual) bent teams up with someone of an "artisan" (operational) bent, and the depth of unspoken, informal operational knowledge that only the craftsperson possesses is extracted. Once this knowledge can be articulated into conceptual terms, it can be merged and combined with other knowledge, likely even triggering the creation of new conceptual knowledge in a very powerful form of learning. However, this cycle is truly effective only if that knowledge is then brought back to bear on solving operational business problems in a new way. To do this requires some form of experimentation in an operational environment (lab or field).

To come up with far-reaching innovation, this is likely only the first iteration of a series of cycles of this type. The intention is to cycle through this process several times, refining the basic concepts and practices discovered in the first iteration. The cycle time needs to be kept relatively short—the more iterations that can be completed in a given time frame, the greater is the power of melding the best available operational knowledge with the best available conceptual knowledge to create breakthrough learning. Once again, this basic learning cycle we have just described is essentially the spiral model of software development advocated by Barry Boehm.[50]

---

[50] Boehm, "A Spiral Model of Software Development and Enhancement."

7.6. IMPROVING HANDOFFS IN THE WATERFALL MODEL

In the last several sections, we have addressed the topic of learning on more of a microscopic level, looking at knowledge transfer and learning on an individual basis, as well as between individuals within an organization. In this section, we turn our attention to transfer of knowledge between groups within an end-to-end business process. Specifically, we will focus on the most common way of describing the stage-by-stage transfer of information performed in the process of building a TMS solution—the waterfall model.

Within the S&NI life cycle, as well as within the TMS life cycle, there are a series of stages that need to be conducted. The sequential steps form what has traditionally been referred to as a waterfall process model. Typically, each stage of the process is delegated to, or at least led by, different departments or groups. These specialized groups interact sequentially with one another, so that the solution to one group's problem at a given stage establishes constraints in terms of which the group at the next stage must resolve its problems. Although it may be obvious, it is important to realize that steps in the waterfall model can be regarded as part of a learning process, and the exchanges between upstream and downstream stages can be regarded as knowledge transfers.

The effectiveness of the overall process is governed by the collective effectiveness of the sum total of all the handoffs within the process. The effectiveness pertains to the speed with which the handoffs occur, as well as the accuracy. An ineffective process can be the result of either slow handoffs, causing a lengthy process, or inaccurate handoffs ("throw-it-over-the-wall"), causing errors in the process. A key motivation for examining the staging of handoffs within the waterfall model is the powerful relationship that exists not only between adjacent stages, but more critically, the end-to-end relationship that exists between the headend (conceptual design) stages of the process and the tail-end (implementation) stages. Most notably, *the later in the process the errors are detected, the more they cost to fix, whereas the earlier in the process a decision is made, the more project resources are committed as a result of the decision.* These relationships have been analyzed and documented extensively.[51]

We have opted in this chapter not to reject the waterfall model as being fundamentally flawed, but rather we have taken it as a basic way of doing business that is well established and understood intuitively by everyone involved. However, improving the way the waterfall model works for defining and building TMS solutions in the S&NI life cycle is paramount to improving the rollout process for new network technology and services. The central need is to enhance the timing and character of knowledge transfer within the waterfall model. The downstream groups need to be incorporated into the learning process earlier—not only those groups immediately downstream of the headend stages, but even groups at or near the tail end of the process.

[51] D. R. Graham, "Incremental Development: Review of Nonmonolithic Life-cycle Development Models," *Information and Software Technology*, Vol. 31, No. 1, January–February 1989. See also Boehm, "A Spiral Model of Software Development and Enhancement."

Tieing the headend and tail-end stages together is significant for several reasons. As mentioned earlier, the sooner a design error can be uncovered, the less costly it is to correct. Operational expertise at the tail-end stages are often best equipped to uncover these flaws that may not be apparent to the conceptual experts in the head-end stages. By bringing in the hands-on expertise earlier, the critical decisions that result in the greatest commitment of project resources can be made with input from those groups that typically inherit responsibility for managing the bulk of the resources committed. The powerful, and at the same time, painful part of this mix is that operational and conceptual knowledge must be transferred. This is especially critical for *complex* systems where the characteristics of the system need to be discovered in the upstream stages, assembled into an architecturally consistent structure, and then communicated effectively to the downstream stages in a conceptually integrated manner.

In Figure 7.7, the traditional, *phased approach* to transacting handoffs between an upstream and a downstream stage of the waterfall model is contrasted with an *overlapping approach*[52] that is consistent with the objectives just outlined.

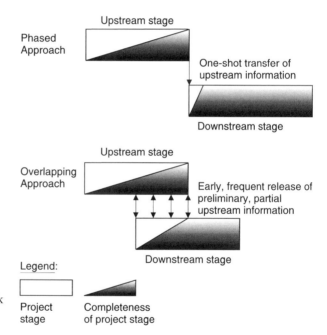

Figure 7.7 NP Service and Network Initiative (S&NI) Life Cycle.

7.6.1. The Phased Approach

In the phased approach, the upstream group proceeds until most of the decisions for that stage have been made. Then the information deemed relevant is "thrown over the wall," or transferred to the downstream group in one large

---

[52] *Dynamic Manufacturing*, p. 314.

chunk. The downstream group begins its stage with all the information provided by the upsteam group. One problem is that they may not fully understand the choices, tradeoffs, and interrelationships involved in the upstream decisions. In addition, the downstream group may face additional constraints beyond those assumed in the upstream design, but potentially conflicting with them. A major weakness of the phased approach is that resolving such conflict requires a feedback mechanism that does not usually exist. Compensatory actions taken by the downstream group tend to compound problems, and so the phased approach usually results in longer development times, poorer use of resources, and lower quality designs.

The upstream group does not always know or understand what information is needed by the downstream group—the more formal the transfer (i.e., a voluminous requirements document), the less subtle or tacit information is transferred. Only by involving the downstream players *actively* in the upstream stage can the deeper insights (which are not so readily communicated in formal media) be effectively transferred to the downstream stages. In the traditional phased approach, key information is missing at critical decision points. Key design decisions that will heavily impact the cost, efficiency, and complexity of routines that will be required at later stages (stages that predominantly involve operational knowledge) are made without the benefit of direct operational knowledge.

7.6.2. The Overlapping Approach

In the overlapping approach, the downstream stage gets involved well before the upstream stage is complete. The two stages are linked by a two-way, continuous stream of information exchange that is often tentative and fragmentary (refer to Figure 7.7). Information moves between upstream and downstream stages in small increments instead of in one large installment. The knowledge transfer is more gradual and sustained compared to the "one-shot" transfer in the phased approach.

The downstream group is allowed to gain insight into problems being wrestled with in the upstream stage, and has advance warning of possible outcomes that may impact the downstream group's work. The downstream group learns to anticipate upstream outcomes, based on the early input they receive. This not only allows them to plan their work better in advance, but it also gives them the opportunity to provide input to upstream decisions. The early input from downstream improves the ultimate quality and effectiveness of the overall project since potential problems are identified earlier in the process, when the designs are much more malleable, and potential design flaws are more readily corrected and less costly to remedy.

The interaction across the boundary between the upstream and downstream stages can be designed as a miniature learning cycle, exchanging operational and conceptual knowledge as described in section 7.5.3.5, and the benefits of the OADI (observe, assess, design, implement) learning cycle described in section 7.4.1.1. On a more macroscopic level, the entire end-to-end waterfall process can be modeled as a learning cycle as well. A key component of making this work effectively is to take advantage of the architecture team to provide continuity across all stages of the waterfall process. We will address this matter in the concluding section of this chapter.

To make the overlapping approach work, the upstream group must be willing to trust the downstream group's ability to cope with ambiguity and uncertainty in subsequent design changes. The upstream group must also be willing to risk releasing preliminary information and to accept feedback from the downstream group that could hamper the upstream group's freedom to make changes later. Since the groups share more of the responsibility for the outcome of their overlapping efforts, deeper commitment is required. An upstream group must make sure its design is right, and it must also see that downstream groups can implement it properly. A downstream group must share responsibility for design and not use "poor upstream design" as an excuse for whatever problems develop downstream.

To be successful, the overlapping approach must facilitate information-sharing between upstream and downstream stages so as to surface potential conflicts as early as possible in the project life cycle. Once surfaced, the conflicts must be resolved in a manner that accommodates the needs of both groups. And instead of escalating conflicts, the groups must strive to resolve issues at the lowest level in the organization. The architecture team can play a key role in managing conflict in the TMS solution architecture process by performing the following:

1. Look out for the good of project solution as a whole.
2. Play an even-handed arbitration role between the various groups involved.
3. Resolve disputes based on global, not local optima.
4. Encourage communication among groups that never knew they needed to have meaningful and sustained exchanges.
5. Go out of their way to make explicit the assumptions that may be thought to be obvious but are likely not fully shared by all participants.
6. Create "singular unity of purpose" through shared mental models: get groups to stand shoulder to shoulder, facing common problems rather than toe-to-toe facing phantom rivals.
7. Create healthy tension; leverage the energy that arises from natural conflicts and contradictions.

7.7. PUTTING IT ALL TOGETHER: THE DESIGN FOR COMPLEXITY

In this chapter, we have addressed the basic question, "What needs to be done to reduce the time it takes to roll out TMS solution support for a new service and network initiative (S&NI)?" We have addressed this question by examining the life cycle for the SN&I process and by attempting to discover where the *essential complexity* arises in this process. More generally, we have looked at this life-cycle process as a specific case of a more general class of complex problems, whose solution requires a complex System (system-of-systems). Because each problem of this type is complex, by definition it "exceeds the capacity of a single individual to understand it sufficiently". As such, these tasks can only be carried out by an architecture team. To define a complex System solution, the architecture team must first adequately define the complex problem and then architect the complex System solution.

As we have seen, the architecture of a complex system is tightly linked to the structure and dynamics of the organization that defines it. As such, the architecture team must of necessity work through the life-cycle process based on the shared mental models, frameworks, and routines that are established within the organization. The architecture team must manage the TMS rollout within the S&NI life cycle as a *learning process*. We have examined models of individual and organizational learning, and have proposed that complex problems require organizational *double-loop learning*. Furthermore, we have seen that to be truly successful, this learning process must tie together two fundamental types of knowledge—operational (know-how) and conceptual (know-why). Finally, we have seen that while the waterfall model is a useful way of structuring the life-cycle process, improvements need to be made in the way information is exchanged between stages in the process.

To conclude this chapter, we will briefly summarize the steps that the architecture team must follow in defining a complex System solution:

1. *Define the problem (structure the problem space).*
   This step primarily involves working with the client to define and refine the scope of what is included (and, just as importantly, not included) in the problem domain.

2. *Discover/create Ashby control knowledge and shared mental models.*
   This step involves the best and brightest talent that can be assembled as part of the architecture team, defining in solid technical detail the fundamental characteristics of the problem domain. This step is at the center of what the architecture team does. Ashby knowledge is created, consisting of the union of two types of system behavior: (a) under normal, intended operating conditions; and (b) under adverse conditions when the system operates in a potentially degraded or partially/fully failed condition, or experiences exception or error conditions in processing. A key product of this step is the initial definition of the shared mental models (SMMs), which capture the essence of Ashby knowledge.

3. *Structure the solution space into solution components.*
   Following the definition of the problem, but essentially in parallel with the creation of Ashby knowledge for the problem, the solution components that will potentially be used to construct the solution need to be sized up. Involved in this process is creating a map of all the component systems that will fit together to form the composite solution System (system-of-systems). It is critical that this process create an unbiased, accurate assessment of the actual capabilities of the component systems. Once the requirements of the problem space are known, this map can be used to perform a gap analysis to determine where the solution components do and do not cover the problem space.

4. *Farm out solution components (allocate problem space to solution components).*
   In the critical step that shifts the project from focusing on the problem definition to developing the solution, the detailed characterization of the problem is allocated to specific work groups responsible for the various solution components. If gaps exist between what the problem requires and what the current set of solution components provide, then these gaps need to be assigned to new or existing work groups. In order for the architectural integrity of the project to be maintained across this transfer, the SMMs that embody Ashby knowledge must be clearly articulated to the solution groups. The solution groups are much more capable of fully grasping these SMMs if they have been involved in more than a superficial way in the earlier steps.

5. *Communicate the shared mental model (embodying Ashby knowledge) to the solution groups.*
   During the earlier steps, the architecture team is in the process of creating shared mental models that capture the essence of the solution required to solve the problem. The SMMs are the core of the project's architectural integrity. The architecture team needs to begin socializing and communicating the SMMs early and often. The architecture team must have linkages to all stages of the process and must involve key expertise at the appropriate point in the life cycle. Journeys need to be taken further downstream than the traditional life cyclists would ever be comfortable with. The architecture team can begin to ingrain the SMM into the fiber and fabric of the organization—discovering early where the barriers and resistance will arise, and what detailed technical issues need to be worked in what order.

6. *Design first iteration of solution design.*
   Doing what they do best, the solution work groups define a first iteration of their individual solution components.

7. *Cycle solution design against problem definition (Ashby).*
   In this step, the solution components are assembled together (at least on paper) and then analyzed against the characteristic (Ashby) knowledge of the problem. This step is necessary because the solution components (individually and in collection) introduce new problems that need to be integrated into the overall problem domain. After all, these new problems become part of the total problem that must be solved by the solution.

8. *Produce solution components.*
   Assuming all of the previous steps have been followed, this is pretty much business as usual.

We have seen that the key "innovation" that is needed to reduce the time it takes to roll out TMS solution support for a new service and network initiative (S&NI) is the introduction of an architecture team. The architecture team is linked to processes upstream (in the earlier stages of the S&NI life cycle) and to key stakeholders downstream in the TMS life cycle. The architecture team is responsible for getting its collective arms around the complex problem posed by the telecommunications needs of the new service and network initiative. Once the architecture team has defined the problem sufficiently, they are responsible for mapping it to the groups that will produce the solution components. The essential role of the architecture team is to ensure that architectural integrity is preserved throughout the life-cycle process, so that the solution being provided matches the requirements of the problem at hand.

Enrico Bagnasco
CSELT
and
Marina Geymonat
CSELT

Chapter 8

The Impact of Telecommunications in Europe and the Requirement for Network Management

8.1. INTRODUCTION

The telecommunications industry is significant for Europe because of its impact in the techno-economic sphere, in industrial policy, and in broader social context.

At the techno-economic level, a common vision of pervasive broadband communications is emerging as a necessary infrastructure for industrial and social development in all the major industrial and trading areas of the world, that is, the National Information Infrastructure (NII) initiative in the United States, the Beacon Initiative in Canada, the Visual, Intelligent and Personal Network in Japan, and so on. As was indicated in the recent Commission White Paper on Growth, Competitiveness and Employment, today telecommunications industries account for an annual market in terms of services of about US$ 230(B) at a world level and about US$ 69(B) at the Community level. The equipment market is worth about US$ 66(B) at the world level and about US$ 21(B) at the Community level. The expected annual growth rate until the year 2000 is 8 percent for services and 4 percent for the equipment market. It is estimated that this sector alone will account for 6 percent of GDP at the end of the century, not including the indirect effects of network installation and operation on the economy as a whole. In 1992, a report by the Economic Strategy Institute in Washington, D.C., indicated that the accelerated introduction of high-speed communications could boost U.S. economic growth by 4 to 6 percent over the next 15 years. A similar study carried out for the Commission of the European Communities in early 1993 showed that a similar effect could be generated in Europe, but that a strong stimulation of both infrastructure implementation and usage innovation would be needed to match the 6 percent growth increment predicted for the United States.

In terms of European industrial policy, advanced communications technologies and services are crucial for consolidation of the Single Market, for Europe's industrial competitiveness, as well as for its balanced economic development. Difficult economic conditions in Europe at present are forcing businesses to consolidate and restructure their activities. New networks of suppliers and clients are needed; flexible and distributed working methods have become a necessary part of new business operations, and advanced communications are now essential to link businesses and business units together. The competitiveness of European industry depends on its ability to take advantage of advanced communications services within Europe and in all other major markets for European products and services around the world. Not only must businesses themselves reengineer their activities, but they must also seek new strategic alliances. These must increasingly span the European economic area and may need to be global in their marketing potential. Advanced communications development is essential. The pace of new service provision will have a major impact on industrial policy development. New communication system developments will also enable the emergence and growth of new industries. For example, the availability of interactive digital video entertainment services will both require and stimulate a fundamental restructuring of the TV program production, scheduling, and distribution industries. It will allow convergence between the traditional publishing industries (books, newspapers, and magazines), the entertainment and games industries, and the television and cinema industries. Many new job opportunities will be created.

In social terms, advanced communications technologies and services—as well as being a vital link between industries, the services sector, and the market—are also an essential link between peripheral areas and economic centers. As such, they are a prerequisite for social cohesion and cultural development in Europe. Thus, telecommunications development will not only be a basis for innovation and a key to competitiveness, but must also contribute to answering European society's needs. The two major European social challenges are in the areas of employment and cohesion. Advanced communications are an essential part of the answers to both of these challenges. With regard to employment, development and use of advanced communications have done more than any other industrial development to stimulate the emergence of new services and sectors of employment. Better communications helps every company reduce costs and expand its market. Whole new service industries are now emerging in which use of fast telecommunications is an intrinsic part of the business process. In addition, telecommunications and information technologies now offer a new flexibility in the employment market. A recent survey showed that 96 percent of Europeans are unwilling to leave their region or country to seek employment elsewhere in the Community. Advanced communications can bring employment to people, through distance working—"telework." It can also stimulate new flexibility in working hours, particularly for part-time work, and in employer–employee relationships (multiple part-time employment patterns).

On the challenge of European cohesion, a recent analysis of economic growth stimulus from advanced communications use has shown that the boost to business competitiveness and economic growth will be strongest in the more industrially developed regions of Europe unless specific measures are taken to promote the decentralization of business services throughout the community.

Among the different ongoing initiatives along these lines of telecommunication development, it is worth mentioning ACTS and EURESCOM. In general terms, ACTS is a research program sponsored by the European Union, open to all industry segments (e.g., manufacturing, network operators, university) and focused on mid-term broadband networks and applications. EURESCOM is a research institute founded by European network operators and covers short-term needs (e.g., network management, access network, strategic studies). Both initiatives will be fully explained later in this chapter.

8.2. THE PAST AND PRESENT NETWORK MANAGEMENT SCENARIO IN EUROPE

The telecommunications sector in Europe has been traditionally dominated by monopolies with exclusive rights of service provisioning in well-identified areas. This approach has influenced strategies as well as the evolution of networks and telecommunications operations. The liberalization process began about four or five years ago, and, it has been steadily accelerating. Liberalization has produced the privatization of state-owned telecommunications companies, the opening of markets to competitors, and the appearance of new players on the scene, for example, the HERMES consortium of railroad companies providing long-distance services—all resulting in more pressure in terms of time-to-market, quality of service, and low cost for traditional players. This market scenario combined with the development of telecommunications networks from wholly analog through the integrated and intelligent digital networks of today to the broadband environment of tomorrow, and the explosion of new services being offered or planned, has stressed the problem of efficient administering, operating, and maintaining of networks and services. So, if transmission and switching equipment have been traditionally considered the main components of telecommunications networks, network operation and management systems have now become as important as the first two.

The history of network management in Europe is probably very typical and similar to that of other regions characterized by the same market environment. Although network operation and management has always been present, it has never been seen as integrated in the network itself. In the era of electromechanical technology, network operation and management was primarily manual with distributed control. With the introduction of computer control into the network, in the form of electronic switches and computer-based operations systems, the concepts of mechanization and centralization were introduced and established. In today's era of digitalization, where the trend is integration of functions, services, and networks, network operation and management has to fully support new network infrastructure as well as new services.

In order to have a complete scenario on "network and service management systems" in Europe, many surveys have been conducted. Overlooking the European network management scenario, the primary aspect to be outlined is that this sector is composed of a large set of varied systems whose status is strongly dependent on that of the managed network. Older and larger networks (e.g., PSTN) have a complex variety of management systems. Basically, two broad classes of management systems are in use today in the public environment: manufacturer products and "home-made" systems. The first type of management systems tends to be installed closer to the Network Elements (NEs) and is tied to the manufacturer's technology. They usually provide a large set of functionalities within the areas of performance and fault management and general data collection for further processing. The scope of these systems tends to be very focused on a defined part of the whole network (e.g., digital switching). The second type of system tends to be installed in the higher layers of the management hierarchy, usually providing solutions tied to the operator's organization. They are dedicated to a restricted set of functions like billing, planning, and overall performance and quality-of-service monitoring. The scope of these systems tends to be wide, covering the whole network. In addition, open technical solutions are limited both by the manufacturers' proprietary technology and by the operator's organizational structure. The lack of complete management standards (information models are only recently becoming available) identifies few of the examined systems as "open systems."

With respect to level of integration, there is a clear predominance of specialized systems developed for and tailored to specific needs. This results in a poor level of integration among these systems and in inconsistencies and overlapping among the management data stored in the databases. Another general aspect is the extensive involvement of human operators in many management aspects, so that the level of automation in some areas is quite low. This is an important point because a high level of automation will be crucial for the cost-effective management of the complex network environment of the (near) future.

With regard to level of interworking between different management systems, the situation appears to be less integrated in those countries where there have been different operators, either operating in a particular region of the country or in a particular segment of the network (e.g., domestic, intercontinental, international). In these countries, the trend has been toward the development of proprietary support systems. The few exceptions are systems within the accounting management area (e.g., shared billing centers). In general, the level of interworking among management systems of different countries appears to be quite low, again restricted only to some procedures within the accounting management area and to the management of international circuits. Finally, communication among different systems is often affected by physical exchange of paper files, therefore preventing the easy further processing of management data.

8.3. TMN IN EUROPE

With this scenario in place, TMN has been considered a new approach to network management. In particular, of the many proposals identified in the TMN framework, two are the most closely considered in Europe:

- The X-interface, providing an open and standardized interconnection between OSs of different jurisdictions (key for the management of international services in the fragmented European scenario).
- The Q3 interface at the network management level, providing a common denominator to different technologies installed in typical European multivendor networks.

In particular, the X-interface is very relevant in many European initiatives (e.g., Eurescom) which are described later in this chapter.

In order to complete this overview of TMN in Europe, we will identify a set of evolution drivers. Not all of them are specific to Europe, but are probably rather common to different other regions.

8.3.1. Increase of Liberalization and Deregulation

The evolving European scenario has to take into account the Commission of the European Communities (CEC) proposals for the deregulation of telecommunications presented in the "Green Paper," and, in particular, the concept of Open Network Provision (ONP). The requirements that ONP will put on TMN are under consideration for the definition of an open, standard interface to the OSs involved in the management of reserved services[1] where ONP conditions could apply. A key date for deregulation of European telecommunications as well as the management area is 1998. Important aspects like the "nondiscrimination requirement" of ONP mentioned above have to be taken into account. The market will need to establish equality between different actors (e.g., traditional operators, service providers, new entrants) in offering management services with the purpose of avoiding dominant positions in the telecommunications market. TMN will be in support of this driver, providing the reference architecture to define and adopt open interfaces between separate different management layers (e.g., the Service Management Layer and Network Management Layer).

8.3.2. Increase of Automation

One of the most significant differences between today's systems and the TMN environment will be the degree of automation in handling management procedures, functions, and information. The low automation level will gradually be transformed to highly automated processes according to the object-oriented, information-driven paradigm. TMN automation is crucial to providing cost-optimum solutions, thriving in a competitive environment, coping with the growing network and service complexity, adapting flexibly to rapidly changing customer demands, and overcoming scarce OAM&P personnel skilled in the new technologies being introduced (e.g., SDH, ATM).

---

[1] A service provided within a regulatory environment (e.g., license).

8.3.3. Introduction of Cooperative Management

Among the key factors that will influence the near-future European telecom scenario, the most relevant are the implementation of pan-European networks, deregulation of the telecommunications market, adoption of concepts like One-Stop-Shopping, and increased subscriber mobility. These new types of scenarios will outline a number of constraints on management systems, create new demands for accounting services, lead to an increase of information exchanged between operators and service providers, and increase demand for network management systems that can operate across more than one management domain. Cooperative management will then become key. The management information to be exchanged between different actors in the pan-European telecommunications market soon assume a new dimension. It will probably be necessary for European Public Network Operators (PNOs) to allow other PNOs and service providers to access a defined set of their own management facilities. The most probable scenario will rely on cooperation between management systems in different domains, while, in the long term, some long-distance international (e.g., resulting from the ongoing merging process) service providers may appear.

8.3.4. Service Management and Customer Control

Service management is becoming more important owing to the evolution of market pressure and regulatory environment. Competition between service providers will grow. This calls for wider scope of TMN systems, addressing the service management level. The key challenge in Europe, still to be resolved, is the management of pan-European services running over interconnected (and separately managed) national networks. In this near future scenario, there will also be competitive pressure to allow for a greater degree of user control over network resources. Management services will be sold to customers together with network services. This openness to customers will have to be carefully managed. The user's ability to investigate (e.g., alternative networking arrangements) could introduce conflicts. In particular, database management requires careful coordination throughout the shared, multinetwork environment as many activities interact (e.g., reconfiguration following failure versus introduction of new services).

8.3.5. Security of Management

TMN security is becoming more and more relevant with the increased openness of the network environment of tomorrow, demanding a homogeneous approach to security at various levels. To ensure the security and integrity of the entire TMN, all subnetworks and constituent systems must adhere to a common security strategy. Among key security objectives identified for TMN are its integrity, availability of data and functionalities, limitation of damage effects due to violations, accountability of management actions, and confidentiality of information.

8.3.6. Standard Management Interfaces on Future Technology

A large variety of technology is now mature, and it's being introduced extensively in networks (e.g., SDH, ATM). These new technologies have been developed and specified considering the management aspects and, in particular, the standard TMN approach. Their introduction will provide a strong impulse to the overall migration toward standard management systems.

8.4. INITIATIVES SPONSORED BY THE EUROPEAN COMMISSION

Advanced communications technologies and services are crucial for consolidation of the internal market, for Europe's industrial competitiveness, and for balanced economic development. Services are a vital link between industry, the services sector, and the market, as well as between peripheral areas and economic centers. They are also a prerequisite for social cohesion and cultural development. All these considerations have been important concerns of European policy for many years.

Intervention of the Commission in telecommunications also arises from the truly international nature of the industry. Increasingly, telecommunications in Europe is across national borders, particularly for larger companies that are most frequently the pioneering users of innovative services. This implies that individual national actions, be they by governments in the regulatory field or by national telecom operators in the field of services, though absolutely necessary, will not be sufficient. A further level of international actions and coordination is required. This is supplied by the European Commission.

Sector actors have been working together since 1984 under the auspices of the European Union to formulate and implement a coherent European telecom strategy. Over that time, policy has evolved making use of different instruments as appropriate, from Research and Technological Development (RTD) collaboration for the preparation of longer term objectives to the use of quick-acting consortia to address short-term problems. The Commission is expected to continue to make use of one or several of the following methods of work, as appropriate:

- Consortia
- Policy integration
- Mobilization of market forces
- Studies
- Recommendations
- Regulations
- Public interest projects
- Trials
- RD&E collaboration

The Commission's actions in telecommunications are and will be governed by the principle of subsidiarity, which, simply put, is that Commission intervention is made only at a level where a pan-national action is appropriate and necessary, and not otherwise. Areas where national action would supply the most appropriate means is left to that national action.

The EU has at its disposal several policy instruments that can be used, individually or in combination, to reinforce efforts on the level of member states, according to the principle of subsidiarity. These include Growth Initiative, the European Investment Bank (EIB), Structural Funds, Trans-European Networks (TEN), Social Policy, Telecom Policy, and its RTD Policy (Framework Program). In addition, close collaboration with other organizations adds strength to EU-level actions. Specifically, this applies to standardization bodies (ETSI, CEN/CENELEC), specification bodies (EURESCOM), EUREKA, and others.

As reported in the ACTS workplan, in the Commission's opinion Europe's research and industrial base suffers from a series of weaknesses:

- *Level of resources*—In 1991, RTD spending represented only 2 percent of GDP in the EU, compared with 2.8 percent in the United States and Japan.
- *Coordination of research* —Historically, there has been a lack of coordination between the national research policies.
- *Application of research results*—The greatest weakness of Europe's research base is its comparatively limited capacity to convert scientific breakthroughs and technical achievements into industrial and commercial successes.

Following is an overview of the most relevant initiatives focused on the telecommunications sector sponsored by EC, together with a table to help locate, in time, the various projects. The dashed lines represent the organizations, while the rectangles represent the single projects.

8.4.1. RACE

RACE (Research and Development in Advanced Communications in Europe) is a program (1988–1994) of precompetitive research and development that was initiated by the European Commission in 1988, following extensive consultation with the industrial sectors concerned as well as national governments. The main objective of RACE is "to prepare for the introduction of IBC (Integrated Broadband Communications), taking into account the evolving ISDN and national introduction strategies, progressing to Community-wide services by 1995." Within RACE, a number of projects have been focused on TMN aspects, covering issues like functional reference model, reference configuration and architecture, together with experimental implementations focused on traffic management, maintenance, and customer administration. Other TMN projects covered themes like service management, user access to management facilities, mobility management, and management service creation.

TABLE 8.1 EC SPONSORED INITIATIVES

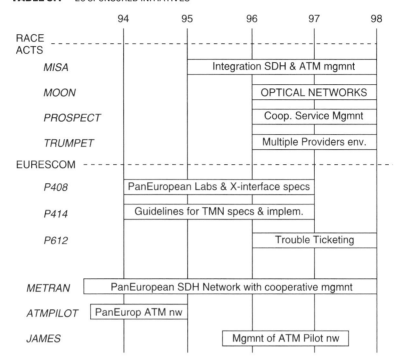

8.4.2. ACTS

The ACTS Program (Advanced Communications Technologies, and Services) represents the European Commission's major effort to support precompetitive RTD in the context of trials in the field of telecommunications during the period of the Fourth Framework Program of scientific research and development (1994–1998).

Intensive consultations carried out with sector actors and other interested parties identify the need for the ACTS Program to concentrate on interworking, integration, and verification through projects on high-speed, photonic, and mobile communication systems, and the distribution of network and service intelligence.

In the White Paper on Growth, Competitiveness, and Employment, the Commission has proposed that member states of the European Union and European institutions should together focus on five priorities:

- Promotion of the use of information technologies, particularly in the public sector, but also through promotion of teleworking.
- Promotion of investment in basic trans-European services, for ISDN and high-speed networking.

- Creation of an appropriate regulatory framework to ensure competition and guarantee universal service and security of information and communication systems.
- Development of training in new technologies.
- Increased industrial and technological performance in European business, notably through increasing research and technology development.

Specifically, in the area of trans-European telecommunications networks, we can identify nine priorities for investment stimulation:

- Establishment of high-speed communications networks.
- Consolidation of integrated services digital networks across Europe.
- Consolidation of systems for electronic access to information.
- Development of European electronic mail services.
- Implementation of interactive video services, based on CATV and telecommunications infrastructures and technologies.
- Stimulation of teleworking.
- Stimulation of telematics links between administrations.
- Development of teletraining services.
- Development of telemedicine services and networks.

Many of these issues are taken up in the ACTS Program.

It is estimated that a total investment in telecommunications of over 500 billion ECU will be required in Europe by 2005—from network operators in infrastructure development, from service providers in developing new service provision capabilities, and from businesses to allow them to obtain competitive advantage from new infrastructures and services. However, the economic and employment advantages of this investment will be realized only if key European infrastructure and service developments are technically compatible, operationally interconnected, and synchronous in time. The ACTS Program has a significant role to play in achieving this goal.

With regard to TMN, following is a brief overview of some of the most relevant ACTS projects: MISA, MOON, PROSPECT, TRUMPET, and WOTAN. All of these projects were started recently, so they will be presented more in terms of plans and objectives than of results.

8.4.2.1. MISA. MISA, standing for Management of Integrated SDH and ATM networks, was born at the end of 1995 and was intended to last three years (i.e., until 1998). Its keyword is "integration," its overall goal being the realization and validation via field trials of the optimum integrated end-to-end management of hybrid SDH and ATM networks, in the framework of the ONP environment. The overall goal is the realization of a multiprovider, multidomain cooperative management in an ONP Environment for SDH/ATM Networks.

Its main objectives are as follows:

- The definition of a new TMN-compliant management service, called Global Broadband Connectivity Management (GBCM), enabling the provision of an IBC (Integrated Broadband Communication) connection-based service on multidomain networks.
- The development, provision, and validation of an integrated and optimum end-to-end management system for etherogeneous SDH and ATM networks, focusing on the Network Management Layer of the TMN architecture.
- The execution of a field trial using European ATM/SDH backbone and access network infrastructure, bringing IBC services to both residential and business users. A small set of European National Hosts (up to six) will be selected, and, subject to negotiation and contracts, some of them will be included in the MISA trial and whenever possible interconnected. The interconnection will take place both at the network (SDH + ATM) and at the management (X.25 or IP) layer.

To obtain end-to-end connectivity, dealing with the multidomain environment, cooperation among the various providers is required. MISA seeks to enable this pan-European connectivity by developing a distributed cooperative management system that can span across the domains. This will offer a one-stop, end-to-end IBCN bearer service to the customer that will be able to access it via an X-user interface, based on the work done in the EURESCOM P408 project (section 8.5.1.2). MISA customers may be end-users who wish to operate their own network services over the IBCN, or they may be value-added service providers who offer managed network service, Virtual Private Network (VPN), and so on, to their customers.

8.4.2.2. MOON. MOON—Management Of Optical Networks—was established in the first months of 1996 in order to define and experiment with a management system for optical networks. Its objectives are the following:

- To identify network elements of the optical network, such as optical cross connects (OCC), optical regenerators, and optical terminal multiplexers (OTMX) which need to be addressed in the context of network management and OAM.
- To examine the applicability of existing TMN and OAM concepts and network protection/restoration methodologies for optical transport networks.
- To adapt or develop OAM, TMN, and network protection/restoration concepts for optical transport networks, including protocols, managed objects, and their defined attributes.
- To demonstrate and validate these concepts in a field trial network.

The field trial equipment of an existing ACTS project (PHOTON) will be upgraded to demonstrate the management of several optical network nodes. The project also deals with network issues: administration, operation and maintenance (OAM), network operation by means of a telecommunication management network operating system (TMN-OS), network applications, and cooperation with national optical transport network in the core area of the European integrated broadband communications network (IBCN).

The core transport part of future broadband communication networks will be built as a fully optical network with a flat hierarchy interconnecting the access network islands where the subscriber access is based on SDH, PDH, or cell-based ATM links. The TDM principles of ATM, PDH, or SDH standards will be confined to the electrical network, for they cannot simply be transferred to the optical domain. Transparent optical Gbit/s links and FDM principles will be used to exploit the benefits of optical routing techniques.

Management of optical networks is an important building block for the further success of the photonic transport network. This network management must be based on OAM concepts and cooperate with the already existing management of SDH/PDH and ATM layers and the TMN used for core networks.

The acceptance of optical network will depend on the OAM and network management functionality it offers. The network provider is interested in making optimum use of the resources. High utilization and low downtime of connections are important because of the huge information content transported in the very high bit-rate streams of the optical network. Compatibility with the existing TMN standards is also a must, for the existing core transport networks are already managed with the aid of TMN. The introduction of the optical network will presumably occur step by step, and photonic network elements will have to be connected to the existing TMN systems. They therefore have to provide standardized interfaces and interworking capabilities.

8.4.2.3. *PROSPECT.* A Prospect of Multidomain Management in the Expected Open Services Market started at the beginning of 1996 with the following main objectives:

- To realize and validate, via a commercial/business end-user trail, the integrated end-to-end management of IBC services, in a multiservice provider and broadband network environment.
- To provide a management architecture for the creation, deployment, cooperative provisioning, and usage of IBC services within a TMN framework.
- To gather commercial end-user and service provider requirements and experiences in using IBC services and to evaluate the ability of the integrated end-to-end management system to meet their needs.
- To empower the end-users to manage their services by providing user access to the end-to-end service management infrastructure.

PROSPECT proposes to provide an integrated end-to-end IBC service management system supporting multiple coexisting IBC services; this means setting up and control:

- Multimedia teleservice offerings from several different providers, including service configuration and status monitoring of client and server systems.
- Interdomain management interfaces between service providers and network operators, to negotiate resource and bandwidth allocations.
- Interdomain management interfaces between service providers and end-users, to agree on service profiles and to monitor service usage and quality-of-service contracts.
- A management system to support operation of a pan-European Personnel Communication Service (PCS).

The technical approach is based on executing a trial that can demonstrate and validate the management of both cooperating and competing services, in support of the service requirements of commercial/business end-users. The level of ambition and complexity of the goal are such that the project has designed a two-phase approach to realization.

In the first phase, the emphasis will be on integrating the network, service, and end-user infrastructure at the four pan-European sites, together with the provision and management of tele-education service to business end-users at the four sites. In the second phase, the trial will be expanded to support mobile end-users: they will move between the four trial sites, and this mobility will be supported by a Personnel Communication Service. More sophisticated management capability will be added to the trial to support the end-user management requirements and make use of distributed object techniques (e.g., CORBA).

The key issues of the project are as follows:

- Cooperative service management in a multiple-service provider environment.
- Multiservice provider management model specification and demonstration of interdomain management, supporting end-to-end service management.
- Management of Personnel Communication Services in a pan-European open service environment.

As a conclusion, this project will provide solutions for the design and implementation of cooperative services in a pan-European open service market, as well as for the design and implementation of management systems to support the operation of such services. Moreover, it will provide valuable insights into customer requirements for the operation and control of customer broadband services

8.4.2.4. TRUMPET. The Global Information Infrastructure (GII) will feature a multiplicity of network providers, service providers, as well as sophisticated users, all with their own management systems. To ensure that these management systems can interoperate in the respect of the rights and responsibilities of the actors concerned, an interdomain integrity policy is required. TMN's Regulators, Users, and Multiple Providers EnvironmenT-TRUMPET aims at developing a mechanism to ensure the required integrity of interdomain access to the TMN-based management system. Access in this context includes both inter-TMN access and user-to-TMN access. Constituent objectives are as follows.

- To propose a policy as well as techniques for ensuring communication integrity in an interdomain, multiprovider, management environment, including user access to management.
- To design and construct these mechanisms within scenario environments.
- To validate the proposals and mechanisms through two trials involving TMN domains in National Hosts and real users.
- To disseminate results to a wide audience and to influence the ongoing standards definition by contributions to standardization bodies.

An overall concept and integrity policy will therefore be established for interdomain TMN communications, taking into account work done in RACE projects, regulatory policies (ONP), and the status of work in standardization bodies (ITU-T, ETSI, ISO) and forums (ATM, OSI and NM, TINA-C). It will define the information that should be exchanged and the integrity requirements for such exchanges. The overall system design will take into account the needs for practical implementation and will be scaleable to large public networks. ODP concepts and techniques will be used to structure the work of the project. Design and prototype construction will draw heavily on the output of RACE projects.

The trial environment will use the National Hosts together with other networks. Real users and applications will be involved, and a number of scenarios will be specified in detail in order to ensure that trials provide a comprehensive validation of the proposed solutions.

The key issues are:

- Integrity requirement and solution for interdomain TMN communication.
- Demonstration of solutions in real environments.
- Contributions to standardization bodies, concerning TMN and the application of ODP.
- Guidelines for the exploitation of results by users, network operators, service providers and system suppliers.

This project will therefore propose solutions that will mediate the requirements of external users, network operators, and service providers for access to each other's management systems, while respecting system integrity, confidentiality, and means of auditing contractual negotiations.

8.4.2.5. WOTAN.

The aim of WOTAN (Wavelength-agile Optical Transport and Access Network), is to provide an end-to-end wavelength-agile optical communication system for public telecommunication networks. This agility increases the bandwidth capacity of the network and enhances traffic management flexibility. WOTAN will investigate the optimum mix of optical and electronic solutions for switching and networking, so as to maximize the evolutionary potential of the network.

A major objective of the project is to demonstrate how core and access networks can be effectively integrated through the common use of WDM (Wavelength Division Multiplexing) channels. Moreover, with the use of wavelength agility, different PONs (Passive Optical Networks) may be connected simply through the choice of a wavelength. Extension and evolution of such techniques may enable considerable reduction in the number of switches necessary in a network.

The truly end-to-end nature of the WOTAN project will address these main issues:

- Interworking at the access and core network interface.
- The role of an all-optical end-to-end link across the access and core network.
- Managing wavelength-agile PON access networks and wavelength selective core networks within a unified networking environment.
- Interworking of management platforms for the optical access, optical core, and electrical (ATM/SDH) network fabric.
- The impact of customer bandwidth demands and applications on core and access network design and operation through interaction with other ACTS projects and National Hosts.
- Contribution to the international standards debate in the area of wavelength standardization.

The project will therefore bring an increased understanding of how best to construct, control, manage and evolve toward a lean, yet future, proofed (i.e., scaleable), end-to-end broadband national network platform. Strategies for implementing WDM technology in a public telecommunications environment are another expected output of the project.

8.5. INITIATIVES SPONSORED BY EUROPEAN OPERATORS

8.5.1 EURESCOM

The European Institute for Research and Strategic Studies in Telecommunications (EURESCOM) was founded by 20 public network operators from 16 European countries on March 14, 1991 in Heidelberg, Germany. At the beginning of 1995, the Institute had, in practical terms, 24 active shareholders from 21 European countries.

With the world of telecommunications becoming increasingly competitive and R&D resources more scarce, public network operators saw the need to form an organization for precompetitive R&D. In this organization, the shareholders play different roles as initiators of different R&D projects, as performers of these projects, and as receivers of the results from the projects. This cooperation is considered necessary to establish Europeanwide telecommunications services, in addition to existing telephone services. Among the many reasons for forming EURESCOM are:

- The new regulatory environment in most of the European countries.
- The increase of required resources for R&D in the creation, deployment, and operation of the new generation of telecom networks, equipment, and services.
- The need to ensure the interoperability of service and to promote the development of Europeanwide services.
- The results of the European research programs, particularly the RACE programs presented above.
- The establishment of ETSI as an open forum for standardization.

In general, EURESCOM supports shareholders in order to establish Europeanwide telecommunications networks and services. The goal is to make these services available all over Europe and to see them function in the same way everywhere independent of the network operator who provides them. This means that EURESCOM has to work with the services themselves, the networks that provide these services, and the platforms and technologies that support these services and networks. In addition, EURESCOM has to carry out studies in order to be able to recommend longer term strategic options concerning R&D in services, networks, and supporting technologies. More precisely, EURESCOM's objectives are to:

- Enable the development of harmonized strategies via strategic studies.
- Stimulate and carry out prenormative R&D projects.
- Specify harmonized telecommunication networks and services.
- Stimulate and technically support field trials to be carried out by PNOs.
- Contribute to European and worldwide standardization.

EURESCOM currently has projects in the following work areas:

- Strategic studies
- Telecommunications services
- Intelligent network
- Telecommunications management network
- Infrastructures and switched networks
- Software requirements and practices

TMN activities started in 1991, with a pre-study aimed at identifying the specific area of work for EURESCOM, taking into account existing results and ongoing efforts at both the International Standardization Organizations and in RACE programs to take maximum advantage from them and to avoid duplication of effort and divergence from current trends. The key difference in the EURESCOM target, with respect to that of the aforementioned organizations, is that EURESCOM also addresses organizational issues (how shall European PNOs organize themselves with respect to each other in order to provide pan-European services and their management) and implementation decisions.

With respect to the management organization, three possible scenarios are being explored:

- *A joint management*: a cooperative venture of two or more operators and providers under one single jurisdiction; such a venture operates as one legal, economic, and operative unit.

- *A cooperative management*: two or more parties (operators, service providers, customers) set up a contractual agreement among themselves on how they are going to make available/use other parties facilities to manage specific services. The use of resources may be subject to charge on a per-use basis.

- *A pan-European management*: a service is set up for the whole of Europe with a single body as the legal, economic, and operative entity to provide and operate this service. European transnational regulations do apply.

It is felt that different scenarios will coexist in Europe, each better fitting certain services.

Over the years, EURESCOM has been involved in a number of TMN-related activities such as

- Intelligent Network (IN) and TMN service testing
- TMN operations system platform
- TMN guidelines and information model
- Management of interconnected SS#7 signaling networks
- TMN management of IN-based services
- Management of pan-European transmission
- Security and integrity
- Joint/cooperative management services definition and evolution
- TMN organizational model

With regard to TMN, following is a brief overview of some of the most relevant EURESCOM projects: P406, P408, P414, and P612.

Among the major problems European public network operators are likely to face in the near future are the realization and the interoperability of TMN systems, and the issues faced in deploying them. Given that one of the major trends in the telecommunications environment is the so-called globalization of the market, inter-operability of management systems encompasses the PNOs' need to develop auto-mated links to other service providers (SP). Within TMN these links are generally referred to as X-interfaces. The projects described in the following all deal with an X-interface of a different kind.

8.5.1.1. P406-TMN Management of Pan-European IN-Based Freephone Service. The project started in April 1994 with eight organizations contributing resources. Since there is a strong need to provide IN-based services on a Europeanwide basis, the project TMN Management of Pan-European IN Based Freephone Service was set up to provide specifications for managing of a pan-European freephone service, based on IN, in a TMN environment. Its goal was to produce full implementation specifications that should be used for field trial(s) by the participants in the project or by other initiatives. The Europeanwide freephone service has been chosen because of its less sensitive commercial aspects, its satisfactory state of national implementation, and, nevertheless, its significant management functions. The project has concentrated on some initial important management areas concerning freephone, and an emphasis has been placed on the definition of an X-interface between PNOs, taking into account the specific management functions identified by each PNO participating in the project.

The final document produced (i.e., "X Interface Specification for pan-European FPH Service Management") contains a full set of implementation specifications for the TMN management of the pan-European IN-based freephone. Its structure includes three Ensembles (i.e., Provisioning, Maintenance, and Accounting Management of PEIN Freephone Service), a part on X-interface specifications application to the existing IFS, another on the Service Level Agreement Template and Security Aspects, and a last one on the PEIN Freephone Service Information Model. These specifications are expected to be the basis for future implementations and experiments, in the PET-lab infrastructure, within the scope of the follow-on project P612.

8.5.1.2. P408—Experiments and Field Trial Support. EURESCOM has produced a large amount of outputs and results in the network management area, and the need to have a Europeanwide infrastructure, allowing field trials, validation, and experiments of the results achieved up to now, has now been felt.

Previous EURESCOM projects have addressed issues related to the cooperative of the future Europeanwide SDH and ATM networks (e.g., METRAN). P408 relies on their output, as it concerns both ATM and SDH cooperative management, supporting a number of trials that will test and validate these results. It has also provided extensions to the available specifications, with particular reference to aspects of the X-interface. Attention has also been given to trial support in the areas of HMI, operator interactions with the management systems, and the area of security management.

The general aims of the P408 project are

- To validate and enrich TMN specifications.
- To get experience with the interdomain management via the X-interface.
- To provide the necessary building blocks for developing future management services.
- To define interconnection agreements and to implement interconnection procedures.

The project consists of two parts:

• Distributed Europeanwide TMN Laboratory setup (PET Lab).
• Validation and verification of specification for the X-interface (for ATM and SDH management), HMI, and security mechanisms.

For the methodology used and more details on the pan-European TMN Lab, together with some conclusions drawn at the project ending, see section 8.7.

8.5.1.3. P414-TMN Guidelines. In the EURESCOM TMN there is a constant need for harmonization. In particular, the following aspects have been selected as important for this work to continue:

• Management information models for inter-PNO TMN interfaces
• Recommendations related to TMN OS platforms
• TMN systems implementation guidelines

The utilization results of harmonization efforts present a challenge and need to be addressed specifically.

During the past three years, EURESCOM has devoted substantial resources to studies and specification activities in the TMN area. This is happening at the same time that many normative bodies (ITU, ETSI, etc.) have been advancing the work on defining architectures, methods, and models for the realization of network management information exchanges on the basis of open networks. Ideally, its results in the TMN area should form a coherent set that is consistent with the work of the normative bodies. This is difficult to achieve because work in the normative bodies is constantly evolving. Those involved in the development of TMNs are influenced by the evolving work of the normative bodies, and EURESCOM TMN projects are no different. For this reason, EURESCOM promotes activities for the production of guidelines concerning the harmonized development (specification and implementation) of TMN implementations in the context of interworking of PNO-TMNs. The aim is to achieve a common interpretation of standards and draft standards and recommendations, in the TMN area among EURESCOM projects. Because interoperability between PNO-TMNs is a focal concern of EURESCOM, management information models have to be given special attention. Substantial progress has been made with the approaches to collecting, storing and disseminating management information models in EURESCOM P223, as well as providing guidelines in the area of reference point and interface specification methods.

The scope of EURESCOM TMN's area projects is wider than the normative bodies, in the sense that TMN implementation technology is included. PNOs have expressed a keen interest in the OS platforms on which TMN applications will be built and executed. Investigations of technologies, such as OS platforms, have been carried out in the EURESCOM P208 project. Important aspects of this work are the monitoring and influencing of developments in the area of open IT application platforms and the evaluation of options for extending the platform facilities in order to better provide for TMN applications.

As the work progressed, it became evident that a pragmatic dependency existed between platform technologies and the guidelines for development of TMN systems. This is particularly noticeable when the transition is made from requirements specification to implementation. EURESCOM TMN work aims at improving TMN guidelines for the transition from the implementation-independent (specification) to the implementation-dependent perspective; the means to achieve such a transition have to be further investigated.

At a very general level, the P414 objective can be expressed as follows:

- To increase the effectiveness and quality of TMN development activities in the areas of TMN functionality specification and design, common management application platforms, and inter-PNO TMN X-type interfaces—all this through the development, refinement, and dissemination of harmonized guidelines for TMN implementations.

As usability of results is another important issue, the following objective is also included:

- To determine the methods and means to achieve high takeup of the project results and to apply them in the formulation and dissemination of results.

At a more operational level, the main objectives may be placed in four clusters concerned with

- Management information models and requirements specification.
- TMN OS platforms.
- Specification to implementation transition guidelines.
- Interoperability.

8.5.1.4. P612-X-Interface for Trouble Ticketing and Freephone Service Management.
The United States has been witnessing substantial implementation progress among local exchange and long-distance carriers for automated exchange of trouble tickets and carrying out order-handling functions. Within Europe, there is a need to move rapidly into the implementation phase of these specifications and to gain a practical understanding of what it takes to develop, validate, deploy and operate these interservice provider management capabilities in Europe. EURESCOM has successfully set up a distributed TMN laboratory infrastructure (see section 8.7) on which some experiments have been done, related to X-interfaces for ATM and SDH management. This distributed lab can be used to test and validate a trouble-ticketing interface specification, defined relying on the results obtained in other forums, such as the Network Management Forum and the Ordering and Billing Forum. Furthermore, a set of X-interface specifications for

the pan-European IN-based freephone service, addressing the Service Management Layer and including trouble ticketing, are available from EURESCOM P406.

P612, started in February 1996 and planned to be completed at the end of 1997, has the following goals:

- Validate through a real experiment the specifications elaborated by P406.
- Experiment an X-interface for service management, since project P408 only addresses network management issues.
- Define a validated pan-European trouble-ticketing interface. To obtain results that can lead to the definition of a generalized interface and to address real needs, the trouble-ticketing interface specification should be extended to address at least a second service other than freephone. Given the European scenario, the leased line data service has been chosen as the second service category, specifically addressing its realization in ATM technology (e.g., ATM VP). Furthermore, two different X-reference points will be considered, where the interconnection of trouble-ticketing management systems will be defined: the X.coop interface (SP-SP) and the X.user interface (Customer-SP), where the SP role is situated in a PNO domain and a customer role in a private domain.
- Define a validated pan-European ordering/provisioning interface. The first step will be to identify a core set of functionalities that can be used for ordering/provisioning management independent of the underlying service/ network infrastructure. The core specification will then be extended, initially focusing on International Freephone Service, with an intention to broaden the scope to other services (possibly Leased Line Data Services).
- Tailor the X-interface specification for Freephone Service to a short-term scenario, like the one foreseen for 1997, when access to the International Freephone Service through universal freephone numbers will be offered.

This project will therefore provide both trouble ticketing and ordering/provisioning technical specifications and parameters. It will also validate the feasibility of a TMN-based X-interface for International Freephone Service management. These results can then be fed back to standardization bodies, forums, and vendors and also be used for field trials and pilot studies.

8.5.2. METRAN

The project for a Managed European TRAnsmission Network (METRAN) started around May 1991 with a Memorandum of Understanding (MoU) signed by 24 European PNOs, aimed at achieving a better coordination for infrastructure provisioning and operation, better quality transport network, and coordinated introduction and use of new technologies in Europe. SDH and Network Management were addressed. The goal of this collaboration was to simplify and speed up the process of transmission capacity provisioning between PNOs across administrative boundaries in Europe. This project has allowed the PNOs to offer a one-stop end-to-end service to their customers, taking advantage of the reliability and QoS provided by the SDH network, previously unexploited. The actual studies started at the beginning of 1992, and the work was carried out in cooperation with the EURESCOM P109 project, Management of Pan-European Transmission Networks. The studies addressed the cooperative aspects of the network management layer and the specification of X-interfaces between management systems belonging to different PNOs and initially concentrating on SDH networks.

Currently, a pan-European SDH network spanning 24 countries has been set up, and the one stop end-to-end service has become a reality. An SDH X-coop management interface with enhanced functionalities has been defined in the EURESCOM project P408 (see section 8.5.1.2), which can be used in this context. METRAN has also contributed to the extension of ETSI functional and technical standards.

8.5.3. ATM Pilot, James

The European ATM pilot, set up in the framework of a MoU (signed at the end of 1992), includes 16 telecommunications public network operators from 15 countries in Europe. It has resulted in one of the world's largest ATM networks, the so-called ATM Pilot Network. The goals of this experiment as stated in the MoU are:

- To verify and validate ETSI standards, ITU-T recommendations, and EURESCOM specifications proposed for the new ATM technology.
- To identify and develop innovative applications in the broadband domain.
- To analyze the potentialities of the future broadband service market.

Parallel with the European initiative, national initiatives are under implementation. As an example, in June–July 1993 Telecom Italia set up the Italian Pilot ATM Network in its national domain. The current configuration includes four nodes: Turin, Bologna, Rome and Milan. The Milan node is also the international gateway to the European network and is directly connected with the international gateways of Paris, Zurich, and Cologne. Currently, the ATM VP/VC bearer service is provided, based on a 34 Mbit/s access and on a permanent or reserved mode, either periodic or occasional. Many European projects rely on the ATM Pilot Network to conduct experiments on applications for high-speed video communication, cooperative work, CAD/CAM, and so on.

Feedback from network users shows a high degree of satisfaction with the ATM Pilot Network performance, in terms of both available bandwidth and transmission quality. Users have also helped identify areas with room for improvement, many of which have already been introduced, regarding the kind of services provided, some technological solutions, and the management of the network.

The main conclusion we can draw from this experiment is that high-speed networks based on ATM cross connects are mature enough both from the technical point of view and from the customer's business needs. Based on these results, most European operators have already began the commercial phase of the ATM Pilot Network.

This network has many interesting applications, ranging from multimedia e-mail services to work at home to a system for protecting the cultural patrimony to telemedical applications. The European Union and the G7 countries have also proposed ideas for developing the so-called Information Society.

Of course, key to maintaining the effectiveness, reliability, and performance of the network is its management. Each country is supposed to have its own national network management system (which is also studied in the standardization bodies) on which no assumptions were made here. On the contrary, the main focus here is on the cooperative interface needed between administrative domain (i.e., PNOs) management systems, in order to allow effective and efficient management of the whole end to end connection, independent of the number of jurisdictions it spans.

To achieve this interface, a special working group was set up, called the X-coop, whose goal was to define, implement, and test a cooperative management interface between the operation systems of the various national networks, through the CMIP protocol, addressing the configuration management area.

The specification was completed in May 1995 and covered the setup, reserve, and release functionalities for a VP connection involving more than one administrative domain. Each partner subsequently implemented the spec, thus providing feedback to correct errors, ambiguities, and incomplete definitions.

An abstract test suite was then specified and implemented; next, exploiting the pan-European TMN Labs (see section 8.7), each partner ran the tests with all the other X-coop interface implementations, thus providing additional feedback to the spec itself and deeply validating it. The X-coop specification was then forwarded to ETSI, becoming an official standard, and was enhanced and completed with more management areas (fault, performance, etc.).

The whole project has now officially ended, and its results are being used by the JAMES (the Joint ATM Experiment on European Services) project, which began in January 1996. Relying on the specification and implementation experience, it will provide a precommercial global ATM VP bearer service in Europe.

8.6. THE EUROPEAN TELECOMMUNICATIONS STANDARD INSTITUTE (ETSI)

ETSI was established in March 1988, with a broad membership of any country within geographical Europe and several categories: administrations and national standard organizations; public network operators; manufacturers; users; private service providers; research bodies; consultancy companies; and others. Its highest authority is the General Assembly (GA), which deals with administrative matters such as adopting the budget and approving audited accounts. The more technical issues are addressed by the Technical Assembly (TA), which approves technical standards, advises on the work to be undertaken, and sets priorities.

The substantive work for preparing standards is carried out by the Technical Committees (TCs), which answer to the TA and address specific topics. Table 8.2 reports the list of TCs, together with the topic they address.

TABLE 8.2   ETSI TECHNICAL COMMITTEES

NA	Network aspects
TMN	TMN related activities
BT	Business telecommunications
SPS	Signaling protocols and switching
TM	Transmission and multiplexing
TE	Terminal equipment
EE	Equipment engineering
RES	Radio equipment and systems
SMG	Special mobile group
PS	Paging system
SES	Satellite earth stations
ATM	Advanced testing methods
HF	Human factors

In order to expedite the performance of its work, each TC is divided into several Working Groups (WGs) which do the detailed technical work, with the TC functioning more as a management and advisory body.

ETSI produces four main types of documents:

- ETS (the European Telecommunications Standard), which is the official standard of ETSI, published following passage of the approval procedure.

- I-ETS (the Interim European Telecommunications Standard), produced when the subject is not fully stable or proven; it can remain in this category from two to five years, before being either discarded or converted into an ETS.

- TBR (the Technical Basis for Regulations), which is equivalent to an ETS but is used as the basis for a CTR (a document containing compulsory requirements rather than a voluntary status).

- ETR (the ETSI Technical Report), with no technical specification but complementary information about the technical environment relating to standardization issues.

At the end of 1997 the new TC for the TMN activities has been set up, which is going to take over the TMN related work previously carried on in other groups; following is a brief overview of these groups, that is, NA4, NA5, SPS3, and TM2.

In March 1996, NA4, working on network aspects, released a new version of the NA4-3316 document, better known as the GOM (the Generic Object Model). This library of objects is defined using the standard GDMO notation (CCITT, X.720) for the CMIP protocol, which is general enough to be profiled for many different management applications, for both the ATM and SDH technology. Through its wide scope and usefulness it is becoming increasingly popular in the forums and standardization bodies. The purpose of this document is to give a wide choice of objects that people who write specifications can pick up, profile, and reuse in any particular context.

NA5 produced the Q3 between Management System and Network Element (NA5-2210). This group defines messages and information that can be exchanged for monitoring ATM cross connects with no signaling capabilities, independently of the vendor providing them. The Q3 interface abstracts from the low-level details, thus allowing a standard management of multivendor network elements.

NA5 is currently working on document NA5-2212, which defines an X-interface to be used between public network operators in order to manage cooperatively ATM VP connections spanning many different domains. (In Europe, crossing many different countries, each of them has its own PNO.) Eventually, the plan is to cover all five management areas (configuration, fault, performance, accounting, and security), with the first two completed in 1996. The information model for this interface is based heavily on the GOM (NA4-3316). The core of the specification derives from the ATM Pilot X-coop working group where it has been defined, implemented, and tested (see section 8.5.3.). From 1998 on both NA5 and NA4 TMN related activities are taken over by the TMN Technical Committee.

Historically, the SPS3 group has always worked on the specification of the interface between an access network (AN) and the local exchange (LE). In particular, for the narrowband AN, it has completed the standardization of the V5 interface, and it is now concentrating on defining the broadband VB5 interface. The specs for the narrowband V5 have also been approved by the ITU-T.

The group addresses both the network and management aspects (signaling protocol and information model); for management, it is currently developing two recommendations: (1) V interfaces at the digital service node; management interfaces associated with the VB5.1 reference point (DE/SPS-03049) and (2) management interfaces associated with the VB5.2 reference point (DE/SPS-03045). In the context of broadband network management, the SPS3 is also developing the information model for the ATM Switch (DE/SPS-03019). This activity is tightly connected with ETSI NA4, ATM Forum, and ITU-T SG11 Q25.

The TM2 group deals with both access network management and SDH transport network management. For the access network it is developing three recommendations: generic access networks (DE/TM-2227), optical access network (DE/TM-2209), and fixed radio access (DE/TM-2224).

On the SDH side, work is underway at both the equipment (DE/TM-2210) and network level; the transport network (DE/TM-2230) is modeled on ODP principles (explained in DE/TM-2221), coherently with the work being done in ITU-T Q15. A special group is also devoted to the performance management area (DE/TM-2109).

8.7. PAN-EUROPEAN TMN LABORATORIES: THE EURESCOM EXPERIENCE

EURESCOM has proposed and launched projects in the area of TMN, addressing the problem of the pan-European network and service management. Many results have been obtained, including specifications, guidelines, and organizational models, and should now be tested, validated, and enriched by laboratory experiments.

An experimental system has indeed been established to prepare the ground for laboratory validation of present and future TMN specifications for managing pan-European networks and services, and performing field trials for validation of results. P109 has addressed issues related to defining a network of management systems that will manage, in a cooperative way, future pan-European transmission networks (e.g., METRAN). P105 has considered the same aspects in the context of the pan-European ATM network (ATM pilot). Based on that work, a number of trials have started and will be underway in the coming years.

To follow and support this trend, two distributed pan-European TMN laboratories—PET Labs, one for SDH and the other for ATM management applications, have been set up. Each PET Lab initially consisted of four nodes, distributed over four countries, as depicted in Figure 8.1. The SDH labs are shown with stripes, whereas the ATM labs are in white spotted black. (Germany has both an SDH

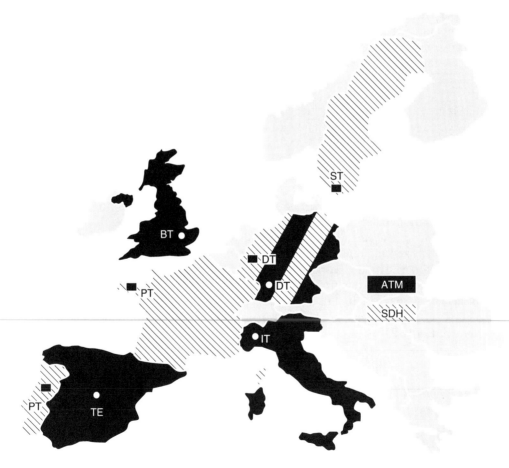

Figure 8.1 PET Lab Nodes.

and an ATM Lab.) The 4 + 4 labs constitute a separate network at the management layer, and each of them manages its own underlying transport network. These two separate layers can be represented as shown in Figure 8.2, where the Management Layer Network is above the Transport Layer Network. This picture is applicable to both ATM and the SDH Labs.

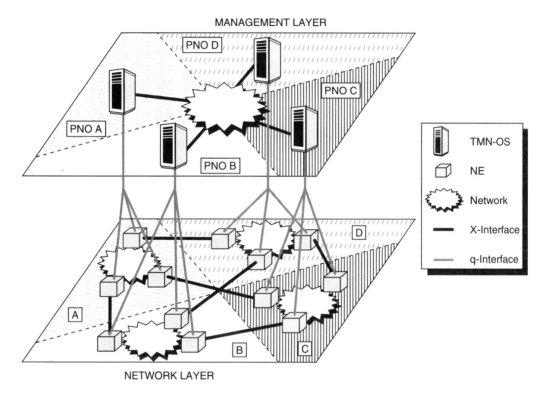

Figure 8.2 PET Lab Management and Network Layers.

All the management systems are linked via the X.25 network at 9600 Kbit/s. This network has proven adequate for exchanging management information. Each laboratory is equipped with an OSI stack implementing the standard CMIP protocol and a platform providing an environment to help develop management applications. As shown in Figure 8.1, different vendors provide stacks and platforms owned by each laboratory. An interesting part of the PET Labs setup has been the protocol-level Interoperability Tests performed in the first phase of the project. A test suite, written in natural language with some GDMO and ASN.1 notation, has been defined and will eventually join the first four, so that protocol-level interoperability can be checked before the actual experiments begin. The tests again come into play any time a new release of the stack software is installed, or when the platform is upgraded, thereby assuring that interoperability will be maintained.

The need to be independent from a single vendor has become stronger and stronger for any network operator. This infrastructure is currently available and needs to be maintained in order to offer two main services:

- A TMN node outside the project can check its protocol-level interoperability with a number of other nodes with different protocol stacks.
- Any node can check its interoperability periodically, for example, in case of failure or of software upgrade.

In the second phase of the project, two X-interfaces between network operators —X-coop for SDH and X-coop for ATM—plus one interface between customer and network operator, X-user for ATM, have been specified and subsequently implemented. A test suite for each of the specs has also been defined and implemented in each lab, according to the methodology defined in the next section. This constituted the first experiment in the PET Lab infrastructure. Every site executed the tests with all the other sites in order to validate the X-interface specifications and their implementations in a real experiment. These are the application-level interoperability tests which, in addition to validating the specs, provide some useful services, notably:

- A new node can, at any time, check its application-level interoperability with all the others, in an ATM or SDH domain.
- Services to check the level of conformance to TMN node security solution (also addressed in the P408 project) are provided.
- Any specification on the CMIP protocol can be implemented and then tested using this infrastructure.

These labs are now being run and maintained in each company by people who have acquired skills and experience in fields that are key to current Europe telecommunications.

The PET Lab infrastructure was represented at Kyoto NOMS'96 by some European partners, demonstrating both protocol and application-level interoperability. Other EURESCOM projects, initiated with the objective of specifying X-interfaces, are planning to rely on the PET Lab for its testing and validation before proposing them to standardization bodies.

The current four ATM labs were expected to grow to seven, as were the existing four SDH labs.

8.7.1. Achieving Interoperability (IOP) Between Multivendor Platforms

Following is an overview of the interoperability testing experience, problems, solutions and results in a multivendor environment, in the context of the EURESCOM P408 project: Pan-European TMN—Experiments and Field Trial Support. The interoperability tests are preceded by a laboratory setup and a test suite specification phase. When the interoperability at the protocol level (ACSE and

CMISE) among all the participating labs has been achieved (first phase), this infra-structure is exploited to validate the X-coop interface specifications.

With regard to the PET Lab setup phase, the complete OSI stack over X.25 has to be configured to interoperate with OSI stacks provided by other vendors. The configuration and parameters setup starts from the lower layers, first experimenting with different configurations with loop-back tests and then testing the pure X.25 connectivity with the other labs.

Next come the higher layers where the same strategy applies: loop-back tests with various configurations and then testing the connectivity at that layer with the remote nodes. The setup phase ends below the application level, which is part of the protocol interoperability testing. Before the protocol level testing starts, all the labs have to exchange the proper platform parameters and application identifiers, such as:

- Network Service Access Point
- Transport Selector
- Session Selector
- Presentation Selector
- Application Entity Title

Once these parameters are set, it is possible to start with the IOP test phase. The following figure describes the test configuration used for the protocol-level IOP.

The Points of Control and Observation (PCOs) are special points through which it is possible to send out (Control) and to read (Observation) the ACSE/CMISE Protocol Data Units to and from the remote node.

TMN Node B depicted in Figure 8.3 will first run the test suite, playing the role of PDU generator and sending the PDUs to the remote Node A, which in turn will have to analyze and check them in reception. Next, the roles are switched,

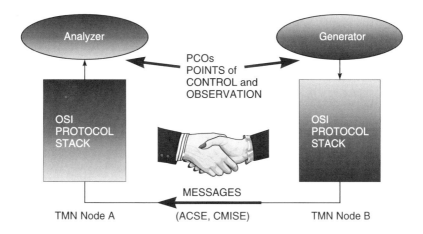

Figure 8.3 Test Suite Configuration.

and the test suite is run again; every couple of labs will therefore run the test suite twice.

The first set of tests specified are aimed at testing the interoperability at the ACSE (Association Control Service Element) level, that is, the capability of setting up an association between two nodes. Even if there are only three possible ACSE primitives and the main one is only the A_Associate, this is the part that causes most interoperability problems. When an association between two protocol stacks is opened, all the layers are actively involved: problems can therefore arise from multiple sources. Another aspect that increases difficulties in interoperating is the fact that most of the parameters contained in an ACSE PDU are automatically filled in with fixed values by the platform itself. Moreover, users are prevented from being able to force the presence or absence of parameters when they are optional (which often happens).

The second set of tests involves communication and consists of an exchange of CMISE Protocol Data Units (PDUs). The PDUs and the order in which they must be sent (and expected) are described in the test suite. This definition is based on a very simple information model, which is needed to define common management information. The syntax and values of the CMIP PDUs parameters have been unambiguously specified making use of ASN.1 and GDMO notation, while the test purpose, the order in which the PDUs are exchanged, and other such information have been specified using natural language. The primitives considered are: Get, Set, Create, Delete, Action, and Event Report. In addition, some tests have been specified to check CMISE scoping and filtering functionalities.

We can summarize the main problems encountered during protocol-level IOP testing, together with the solutions the vendors provided. As already stated, ACSE-level interoperability was particularly elusive. The existence of conditional parameters and their optionality caused two kinds of problems:

1. Platforms that never establish optional parameters could not open associations with those that always request them anyway.
2. Platforms that always fill in the conditional parameters could not connect with those that never accepted them.

In those cases, the second and the first platform, respectively, provided a patch because their behavior was too restrictive.

No platform had problems with CMISE-level interoperability; the only problems at this level are caused by the tools used to compile the ASN.1 and GDMO notation. When complex types are parsed or formatted, the output is not always reliable. The solution to this problem is local and so is of no interest for the multi-vendor environment. Nonetheless, it was a source of considerable problems and delays.

The results achieved in the first phase of the experiment can be summarized as follows.

- A distributed lab for ATM and another for SDH technologies were set up, each of them consisting of four nodes equipped with platforms from different vendors and OSI stacks implementing the CMIP protocol. The labs can be used for field trials and experiments in both technologies, for example, JAMES, METRAN, and ACTS-related projects, described earlier.
- A test suite was defined and implemented, to be used for interoperability testing at the protocol level. This is useful each time a platform or stack software is upgraded and it is necessary to test whether the interoperability with the other labs is maintained. Furthermore, when new laboratories get into play, those tests must be re-run.

In conclusion, we propose some answers for the main issues raised during the project.

Adding new labs: Interoperability does not satisfy the transitive property, that is, if *A* interoperates with *B*, and *B* with *C*, it cannot be assumed that *A* will interoperate with *C*. This brings up the need for any new lab entering a management network to test with ALL the existing labs, at least in theory; in practice, of course, time constraints have to be taken into account.

X.25 link: The choice of using the X.25 national networks to exchange management data across Europe has proved to be a good one. The 9600 bit/sec rate provides real-time responses (less than 3 seconds) to any ACSE and CMISE request. Obviously, more experience will come from experimenting with multiple simultaneous associations, but no particular problem is foreseen even in that case.

8.7.2. Methodology for Specifications Validation: The EURESCOM Example

ATM Pilot and EURESCOM experience has produced a methodology for interface specifications validation, with a CMIP information model. The feedback to the specification comes from various phases with different probability and different kinds of problems detected. The flowchart presented in Figure 8.4 shows how the various phases are carried out and how the feedback is taken into consideration for producing new versions of the specification.

The first step to be performed, when defining the information model for a new specification is to reuse the general-purpose objects defined by the standardization bodies and collected in libraries. After inheriting objects from the standard document, it is necessary to profile them according to the context in which they must be used and according to their purpose. This collection of objects forms the first draft, which must subsequently be compiled by a syntax-checking tool in order to

- Detect all the purely syntactical errors in the formal GDMO and ASN.1 languages.

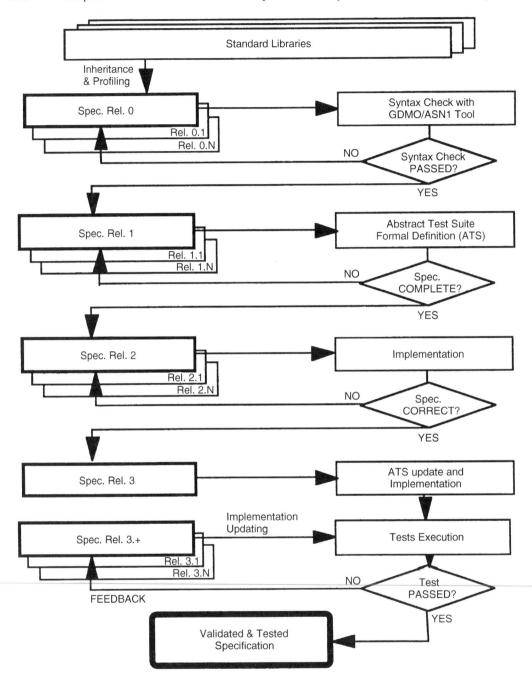

Figure 8.4 Specification Validation Phases Flowchart.

- Detect any inconsistencies within each module (objects used and never defined, duplicated definitions, etc.).
- Cross check the GDMO against the ASN.1 modules in order to completely verify their mutual consistency.

At every iteration, errors are corrected, and at the end of the process, Release 1 is produced.

In the next step, we start defining the Abstract Test Suite that will be used to validate the specification. The test suite can be chosen with different scopes: from an extremely narrow one (focusing on a single functionality and taking into account only the system's valid behavior) to a broad one (covering a whole set of related functionalities and considering some possible invalid behavior).[2] One should not start implementing the specification at this point in time, because errors are likely to be detected during the ATS definition, especially if the tests are written using a formal language (e.g., TTCN). Errors detected in this phase are usually caused by a lack of definition or incompleteness, by situations that have not been taken into account, or by unexpected conditions brought up by a well-thought test purpose. This may cause substantial changes in the specification (represented as Rel. 1.0 . . . 1.N in the figure). The process of writing the ATS, detecting errors of incompleteness, and fixing them in the spec may take quite a long time. When this process ends, the spec implementation can start, based on Release 2.

During the implementation phase, inconsistencies and errors in the spec may be uncovered. This is an iterative process with continuous feedback from Release 2.0 to 2.N. During this iteration, the specification is enhanced, together with its implementation, until no errors are found. At this stage, the ATS can be updated, based on Release 3 of the spec, and subsequently implemented into an Executable Test Suite (ETS).

The last step is the actual execution of the tests. It might be useful here to stress that we are dealing with specifications validation, not software testing. We have therefore assumed that software has been tested locally and that it's working. Errors we are interested in during the execution phase are just those that are due to different interpretations of the spec. Oddly enough, based on the European experience, feedback and enhancements to the specification during this phase are very rare, if a formal testing language has been used. In the EURESCOM and ATM Pilot projects, in which two different approaches (formal and informal) were tried out for test specification, the following results were found: use of natural language for ATS specification requires much less time (and gives less feedback) in the ATS definition phase, but it permits more errors to be carried along to the test execution phase. On the contrary, use of a formal language, though requiring more time for the ATS definition and giving more feedback when passing from Release 1 to Release 2, only rarely leads to errors, due to specification incorrectness.

---

[2] The theory on which the test definitions were based is described in the documents produced by EURESCOM project P201, which specifically addressed this subject.

When the execution phase comes to an end, the specification can be considered correct and complete, for the parts covered by the test definition, and can be sent to the standardization bodies.

8.8. RELATIONSHIPS WITH INTERNATIONAL STANDARDIZATION BODIES

The urgent need for TMN standards is growing rapidly; harmonization and coordination among various bodies producing specifications and standards is particularly important. It is not cost effective to duplicate work, reinvent solutions for the same problems, or produce diverging standards. These are key issues that people working in the normative field are constantly addressing. Nevertheless, monitoring all the ongoing work is not easy, and in some cases this harmonization is not easily achieved. The amount of completed work on which every organization relies when issuing new documents is also a problem: it's not easy to decide to sacrifice backward compatibility (as well as part of the work previously done) to compliance with the work of other bodies. Keeping track of the situation is also difficult, for the picture is continuously and rapidly evolving.

In the previous sections, we emphasized that integration is a key issue even within a single organization: EURESCOM projects have shown how complicated it is to achieve integration among projects addressing different areas but overlapping in some common parts.

Many formal liaisons have been set up between standardization bodies and organizations. In just those related to the European environment, both EURESCOM and ETSI have liaisons with the ATM Forum and ETSI has liaisons with ITU-T for many of its working groups in various areas. In terms of service management, ETSI has liaisons with the NM Forum as does EURESCOM.

Within Europe, each organization tries to keep abreast of projects that are being undertaken by other bodies, in order not to duplicate any work.

ETSI is central to knowledge of what specifications are being produced by European organizations. This "centralization" minimizes the possibility that two groups will be working on the same issues.

Effective cooperation has been carried out, which allows effort and resource savings, but this objective must be continuously pursued. Being aware of what work is going on in other organizations is key to allowing maximum reuse and harmonization and guarding against duplication of results.

References

[1] Cohen, Robert, "The Impact of Broadband Communications on the US Economy and on Competitiveness," Economic Strategy Institute, USA.
[2] Economic Impacts of Advanced Communications; Teknibank, July 1993: Available from DGXIII, Direction B.

[3] ACTS (Advanced Communications Technologies and Services), Workplan, EC DGXIII, August 1994.

[4] Bagnasco, Enrico, 1992, "TMN Evolution," *The Management of Telecommunications Networks*, Ellis Horwood, Chichester, UK, pp. 51–60.

[5] Bagnasco, Enrico, and Roberto Saracco, *TMN in Europe: A Status Report*, INTERCOMM93 proceedings, February 1993.

[6] Callaghan, James, *et al.*, 1992, "TMN Architecture", *The Management of Telecommunications Networks*, Ellis Horwood, Chichester, UK, pp. 23–36.

[7] CSELT Technical Report, Vol. 21, June 1993.

[8] EURESCOM P107, METRAN Technical Network Management, January 1993.

[9] EURESCOM P109, Management of Pan-European Transmission Networks, Deliverable, August 1, 1992.

[10] EURESCOM P201, IN & TMN Service Testing, Deliverable, April 3, 1995.

[11] EURESCOM P408, PanEuropean TMN and Field Trial support, Deliverable, May 1, 1995.

[12] EURESCOM Project Portfolio, January 1995.

[13] NERA, 1992, *Study on the Application of Open Network Provision to Network Management*, NERA/MITA study for CEC–DGXIII.

[14] RACE (Research and Development in Advanced Communications in Europe), Workplan, EC DGXIII, 1991.

[15] Saracco, Roberto, and Bo Rydbeck, 1992, *Towards a pan-European TMN: The EURESCOM project*, 6th RACE TMN Conference Proceedings: invited paper, Madeira, Portugal.

[16] ACTS: European Research and Technological Development, European Commission DG13-B, 1995.

Index

About the Editors

Salah Aidarous received a B.S. and an M.S. in Electrical Engineering from Ain Shams University, Cairo, and a Ph.D. in Electrical Engineering from Louvain University, Belgium. He is currently a senior manager at the Network Management Division, NEC America, working on planning and development of integrated network management systems. Prior to NEC, he was a senior planner at Nortel Technology (formerly BNR), working on a broad range of assignments in telecommunications networks and services planning and development. This includes introduction of new technologies, network and service management requirements specifications, and network management standards.

Dr. Aidarous held various academic positions including associate professor at Aim Shams University and visiting professor at Carleton University, Ottawa, and University of Ottawa. He has been principal investigator for several research grants and contracts in digital mobile radio, CATV, expert systems technology, and process engineering in telecommunications networks.

Dr. Aidarous is senior technical editor of *IEEE Networks* magazine, senior editor of the *Journal on Network and Systems Management*, co-editor and an author of *Telecommunications Management into the 21st Century: Techniques, Standards, Technologies, and Applications* (IEEE Press, 1994), vice-chairman of the Technical Committee on Information Infrastructure (TCII), and is the ComSoc Distinguished Speaker for Network and Service Management.

Thomas Plevyak received a B.S. in Nuclear Engineering from the University of Notre Dame, an M.S. in Nuclear Engineering from the University of Connecticut, a certificate from the Bell Laboratories Communications Development Training (CDT) Program, and an M.S. in Advanced Management from Pace University. Mr. Plevyak is president-elect of the IEEE Communications Society (ComSoc). His two-year term of office begins 1 January 1998. He is the past director of Publications of ComSoc and a former editor-in-chief of *IEEE Communications* magazine. Mr. Plevyak retired after 28 years with Bell Laboratories and AT&T Network Systems (now Lucent Technologies), and is now responsible for operation and network management for Bell Atlantic in Arlington, Virginia. He is vice-chairman of the Inter-American Telecommunications Commission (CITEL), Permanent Consultative Committee 1 (PCC.1), a U.S. State Department appointment. CITEL is a unit of the Organization of American States.

His career spans most aspects of the telecommunications industry, from applied research in direct energy conversion systems to transmission planning, satellite and undersea cable systems, ISDN, video compression, operations systems, and international standards. He holds two U.S. patents.

Mr. Plevyak is co-editor and an author of the book, *Telecommunications Network Management into the 21st Century: Techniques, Standards, Technologies, and Applications* (IEEE Press, 1994). He is author of many technical papers published in a variety of journals, and he was elected IEEE Fellow in 1994 for contributions to the field of network management. In 1995, he received ComSoc's McLellan Award for long-term contributions to the welfare of the Society. Mr. Plevyak will be listed in the *1998 Marquis Who's Who in the Media and Communications*.